Monographs on Endocrinology

Volume 10

Edited by

F. Gross, Heidelberg · M. M. Grumbach, San Francisco · A. Labhart, Zürich
M. B. Lipsett, Bethesda · T. Mann, Cambridge
L. T. Samuels, Salt Lake City · J. Zander, München

W. I. P. Mainwaring

The Mechanism of Action of Androgens

With 12 Figures

Springer-Verlag New York · Heidelberg · Berlin

W. Ian P. Mainwaring
Androgen Physiology Department, Imperial Cancer Research Fund
London WC2A 3 PX/Great Britain

ISBN-13: 978-3-642-88431-3 e-ISBN-13: 978-3-642-88429-0
DOI: 10.1007/ 978-3-642-88429-0

Library of Congress Cataloging in Publication Data. Mainwaring, W.I.P. The mechanism of action of androgens. (Monographs on endocrinology; v. 10). Bibliography: p. Includes index. 1. Androgens. I. Title.
QP572.A5M34 599'.01'927 76–29057

This monograph is dedicated to
Dr. G. F. Marrian, C. B. E., F. R. S.
His work and words largely
forged my interest in steroids.

Preface

My introduction to androgens was neither auspicious nor impressive. I was sitting my *viva voce* examination for a degree in physiology and had haltingly intimated to my examiner (name decorously withheld) that I intended to pursue a career in research. "On what topic?" was the reply. I had been deeply impressed by the work of C. Huggins and C. V. Hodges (Cancer Res. *1,* 293, 1941) on the dramatic arrest of canine prostatic hyperplasia by the administration of stilboestrol. With some enthusiasm, I responded, "On steroid hormones, because I am struck by the profound effects that may be achieved by relatively small numbers of molecules." The examiner sank into deep contemplation before replying, "Young man, have you considered going into teaching?"

Suitably chastened, I finally began my research career investigating the effects of steroids on the nucleic acid metabolism of experimental tumours and on the process of senescence in animal cells. Reaching an impasse in this work, I mentioned one day to Dr. G. F. Marrian that, somewhat surprisingly, we had no understanding of the fundamental mechanism of action of steroid hormones, especially the androgens. He encouraged me to tackle this problem, particularly since exciting new insights were then being made into the interaction of radioactively labelled oestradiol-17β with such tissues as rat uterus. My first faltering experiments on the mechanism of action of androgens began in the late summer of 1968 and have continued ever since, with a particular fillip in 1971 when I was fortunate enough to be made head of a Department of Androgen Physiology. In retrospect, the experience has been extremely rewarding, for after all, I am doing what I initially resolved to do. While at times, the road has been tortuous and beset with deep potholes, at others it has led to clear and exciting heights.

In preparing this monograph, I am deeply obligated to the many people concerned, directly and indirectly. First, to past and present members of my Department, without whose interest, enthusiasm and diligence, nothing could have been accomplished. Second, to the Director of Research of this Institute, Dr. Michael Stoker, C.B.E., F.R.S., for his continued interest in the work and the provision of outstanding research facilities. Third, to my overworked typist and secretary, Cilla Conway, and to Dr. Heinz Götze and his staff at Springer-Verlag for helping to make my transition from random thoughts into lucid words a practical reality. Finally, to my family and friends, for forgiving a few irascible outbursts during the process of writing and giving the encouragement and strength to complete the task.

London, September 1976 — W. I. P. Mainwaring

Contents

Introduction

From the dawning of civilised time, when man forsook the life of the nomadic hunter and adopted the mantle of the communal farmer, his interest in androgens or, more correctly, testicular secretions, has continued unabated. As early man strove to establish his supremacy over the animal world, the extirpation of certain organs from domesticated birds and beasts was widely practised. The capon was recognized as a palatable culinary delight, the ox provided an exceedingly strong but biddable beast of burden and the gelding possessed advantageous features as a dependable yet fast mode of transport. Despite the gradual acquisition by man of an ever-thickening veneer of civilisation and sophistication, these castrated animals remain a feature of life even today. As man developed a more inquisitive and perceptive attitude toward himself, his environment and his procreation, interest in the gonads escalated in the form of complex sexual taboos and fertility rites. In attempts to heighten libido and reproductive prowess, extracts of moulds, plants, and molluscs achieved a measure of mystique as aphrodisiacs and the gonads of slain adversaries were often ingested in the hope of perpetuating sexual activity. Contemporary man tends to place himself above such misguided beliefs, yet for these very reasons the white rhinocerus is nearly extinct and the enigmatic Yeti, the Abominable Snowman of the Himalayas, was claimed to have removed the gonads from its victims. All aspects of this bizarre affair remain to be substantiated, however.

Since the time of the ancients, ablative surgery on male reproductive organs was practised for certain putatively expedient purposes. The barbaric custom of mollifying slaves and captives by castration began in the southern areas of the Aegean and persisted in a wider sphere up to the time of the Inquisition. In Russia, the celibacy of extreme religious sects, notably the Skopetsy, was given an incontestable finality by the removal of the testes of novitiate disciples. Castrati were also common in cathedral choirs until the end of the eighteenth century.

The origins of a more scientific enquiry into the male reproductive glands are hidden in the mists of time but in their authoritative reference treatise, DORFMAN and SHIPLEY (1956) suggest that Aristotle (quoted in *Historia Animalium,* vol. 4, Book 9) was the first to record the profound changes evoked in the cockerel after castration. As with other aspects of cultural and scientific endeavour, further progress and research was lost in the abyss of the Dark Ages, only to begin again in the welcome light of the Renaissance. Then, scientific and aesthetic appreciation of the human body emerged anew. New medical schools were instituted, especially in Italy, where the new concepts of human anatomy laid down by Vesalius in Padua and Pavia brushed away the cobwebs of the past and in the arts, the dignity of the human form reached a pitch of breathless perfection in Michelangelo's statue of David.

These were the trends developing in European cultures, but the remarkably perceptive researches of NEEDHAM and LU (1968) indicate that the Chinese may well have isolated and purified steroids by this time. This is a further example of the inventiveness and skill of the alchemists of ancient Cathay.

Slowly but surely, scientific progress kept pace with this new and invigorating pursuit of knowledge and the following contributions, all acknowledged classics of endocrinological research, laid the foundations of our understanding of the mechanism of action of androgens. Pott, in a paper redolent with the social injustices of his time, remarked on the extremely high incidence of cancer of the scrotum in those desperate unfortunates of the industrial revolution, the chimney boys and noted the impairment of sexual activity associated with this disorder. In one master stroke, the foundations of chemical carcinogenesis and testicular function had been established. Among many contributions from a remarkably colourful and illustrious career, Hunter noted that on removal of the gonads from male animals, many organs distant from the testis atrophied. Hunter surmised that the growth of certain glands, the accessory or secondary sexual organs, was uniquely dependent on the presence of the testis. In a series of bold and imaginative experiments involving the transplantation of the gonads in cockerels, Berthold made a revolutionary contribution. Provided that the blood supply to the gonads was restored, despite their transplantation to other bodily sites, then the reduction in the growth of the comb, so characteristic of the castrated capon, was not evident. However, ligation of the blood supply from the gonads, irrespective of their anatomical location, was manifest in a prompt reduction in comb growth. Taken in concert, Berthold's results indicated that the growth of the secondary sexual glands was maintained by some testicular component that was distributed by the general circulation. Just before the turn of the last century, two enterprising surgeons, Ramm in Norway and White in the United States of America, independently advocated castration for the palliation of human benign prostatic hyperplasia, a disorder of then unproven testicular aetiology. Finally, bringing male reproductive physiology into the present era, Brown-Séquard capitalised on the earlier teachings of the Indian Ayur-Veda and claimed that the injection of testicular extracts, in his case self-administered, was beneficial for the rejuvenation of the ageing human male. All of these historic contributions were brought into close harmony when STARLING (1905) formulated his concept of hormones (in Greek, *ορμαω*, to arouse or excite), the chemical messengers secreted by discrete glands directly into the blood stream in order to regulate the function and growth of organs elsewhere. The unequivocal demonstration of hormonal activity in testicular extracts was first achieved by PÉZARD (1911) and thus he may be considered the father of the contemporary physiology of male sex hormones.

Between 1920 and 1940, impressive advances were made into the physiology and chemistry of hormones, especially those with the basic steroid structure related to pentaperhydrophenanthrene. Many factors were responsible for this success. The preservation and sectioning of tissues was now an advanced skill, being aided by a wealth of precise histochemical stains whereby intracellular components could be specifically located at low concentrations. By such means, the structure of many hormone-producing glands such as the adrenals and ovaries was unequivocally established. Experimental physiology also became a sophisticated discipline and the development of techniques for the selective ablation of endocrine or hormone-producing glands, combined with elegant bioassay procedures, provided plausible explanations for such endocrinologic disorders as cryptorchidism, Addison's disease, virilizing hyperplasias, acromegaly, and diabetes mellitus. Above all, however, this was a golden age of structural stereochemistry and organic synthesis. Through the enterprising application of new methods of organic analysis and identification, the active principles or steroid hormones were isolated from many glands and their precise structures formulated. This was a remarkable step forward and investigators became aware that the biological activity of steroid hormones was a subtle reflection of the nature of their substituent groups and their stereochemical orientation. It also became apparent that steroid hormones fell into separate categories, each essentially related to a restricted biological function. While such a simplistic classification would be somewhat contested today, it nevertheless remains a fundamental tenet of endocrinology, at least in descriptive terms. The testicular steroid hormones were termed androgens (of two-part derivation; the Greek, aner, male and genos, descent) and the principal androgen, testosterone, was first isolated and fully characterised by DAVID et al. (1935).

Beginning about 1950, the general availability of labelled compounds containing relatively stable radioisotopes, first with ^{14}C and then with ^{3}H and ^{32}P, heralded an entirely new and exciting era of biochemistry and cell biology. Rather than describing biochemical and cellular events in general terms, it then became feasible to monitor the synthesis, translocation, and degradation of intracellular constituents with a greater facility and precision than had been possible hitherto. Many research workers in endocrinology were quick to appreciate the new vistas opened up by this revolutionary technology in radioactive tracers, but with the notable exception of establishing the biosynthetic pathway for cholesterol, progress in elucidating the distribution and uptake of steroid hormones was at best confusing, and in reality, extremely disappointing. Many factors contributed to this apparent impasse. First, the labelled steroids and precursors available at that time were of low specific radioactivity and the high doses required to attain working amounts of tracer during experimentation *in vivo* may well have had deleterious effects on rate-limiting metabolic pathways. Second, it gradually became apparent that metabolism of the administered steroid hormones was extensive, thus raising serious difficulties in experimental interpretation. Third, the design of many experiments was open to serious criticism in that labelled steroids were administered to intact animals where important binding mechanisms would already be saturated with endogenous steroid hormones. Fourth, experimenters were slow to appreciate that crucial reactions involving steroid hormones proceeded relatively rapidly and the findings from experiments of long duration were consequently of dubious value. Finally, many investigators erroneously studied the irreversible or covalent linkage of steroid hormones to intracellular constituents; no biological significance can be attached to these studies because endocrinologic events must be freely reversible and thus involve noncovalent bonds only.

It is often difficult to recognise the critical contribution to the evolution of a scientific discipline, but in view of the prevalent confusion regarding the mechanism of action of steroid hormones some 15 years ago, then the plaudits must surely be bestowed on E. V. Jensen and H. I. Jacobsen. Somewhat earlier, Jensen and his collaborators had pioneered the authentic syntheses of tritium-labelled oestradiol-17β and hexoestrol derivatives by introducing tritium gas into the double bonds of suitable precursors in the presence of platinum catalysts. In the now classical paper

(JENSEN and JACOBSEN, 1962) tritiated oestradiol-17β of high specific radioactivity was injected into immature female rats and the distribution of radioactivity was determined in many organs at various time intervals. It was evident that the [^{3}H] steroid was conspicuously retained only in organs sensitive to oestrogens, such as the uterus, and JENSEN and JACOBSEN (1962) concluded that a specific binding mechanism existed for oestrogens in the uterus. Of great importance, the radioactivity of the uterus was recovered as nonmetabolised [^{3}H] oestradiol-17β. In this historical context, two other publications warrant particular mention. EDELMAN et al. (1963) studied the binding of [^{3}H] aldosterone in the urinary bladder of the toad, *Bufo marinus,* and were able to demonstrate by autoradiography that steroid hormones can be bound within the nuclei of steroid-responsive cells. TOFT and GORSKI (1966) introduced the method of sucrose density gradient ultracentrifugation to study the interaction between [^{3}H] steroids and soluble or cytoplasmic subcellular components, an elegant procedure since found to be of almost universal application. The latter studies complement each other and indicate that interactions in both the nuclear and cytoplasmic compartments may be implicated in the overall binding mechanism for steroid hormones.

From such beginnings, subsequent progress in elucidating the mechanism of action of steroid hormones can only be described as dramatic. A specialised monograph of this nature is not a suitable platform for elaborating extensively on the ubiquitous intracellular mechanisms responsible for the high affinity binding of steroid hormones. A comprehensive account of these processes may be found in the reference work by KING and MAINWARING (1974). In brief, steroid hormones pass from the blood by an as yet ill-defined process into all cells. In cells responsive to steroids, the hormone is first bound with a high affinity to a specific cytoplasmic protein or *receptor,* forming a hormone-receptor complex. The resultant complex is then transferred into a restricted number of *acceptor* sites within nuclear chromatin and persists in this nuclear association for a finite period of time only. Once within the nucleus, the receptor complex triggers the transcription of genetic information and other allied processes as required for the manifestation of the hormonal response. Covalent bonds are not implicated in any of these intracellular interactions. The receptor complex is finally degraded by a process that is not currently understood.

In the ensuing chapters, a more critical account will be given of the mechanism of action of androgens. As an inspiration on our way, it is perhaps worth remembering the dictum of Francis Bacon:

> If a man will begin with certainties,
> He shall end in doubts,
> But if he is content to begin with doubts,
> He shall end in certainties.
> (Advancement of Learning, Book I; 1605).

I. A Contemporary Model for the Mechanism of Action of Androgens

There are two ways of learning to swim. Assisted by buoyancy aids, one may enter the shallows and with gradually increasing confidence, finally negotiate the deep water. Alternatively, one may jump boldly into the deep in the belief that a combination of good fortune and a strong preservation instinct will bring the safety of the shore. I encountered such a decision in writing this monograph and after considerable reflection, decided to present a model for the mechanism of action of androgens early in the narrative and to spend ensuing chapters in a more detailed appraisal of its strengths and weaknesses.

1. Definitions

A parlance has evolved in work relating to hormone action and, subject to the present definitions, the following terms will be used extensively.

Target cell: a cell known by classical bioassay procedures to be predominantly under the regulation of a given type of steroid hormone.

Induction: the sequence of metabolic processes through which the synthesis de novo of a macromolecule is accelerated by a steroid hormone.

Deinduction describes the reverse process when the hormonal stimulus is withdrawn.

Receptor: an intracellular component, almost certainly proteinaceous, responsible for the specific and high-affinity binding of a selected steroid hormone and playing an integral part in its mechanism of action.

Acceptor: a nuclear component, responsible for the high-affinity but limited retention of a steroid hormone-receptor protein complex in chromatin.

Of these, only the term *receptor* is open to certain semantic criticism. The receptor concept was first propounded by Paul Ehrlich to provide a feasible mechanism of action of drugs (for a comprehensive review of Ehrlich's contributions, see HIMMELWEIT, 1960). Ehrlich proposed that low molecular weight ligands, such as drugs or even steroids, contained two structural determinants, the haptophore region which combined with the receptor thus permitting the ergophore (or toxiphore) region to initiate the biological response. As far as steroids are concerned, this viewpoint has been modified by subsequent theories of induced-fit and allostery so that the ergophore determinant is now considered part of the receptor molecule rather than the ligand. The mechanism of action of drugs was taken a stage further by CLARK (1933) who suggested that dose-response relationships could be explained by the number of receptor sites and their relative occupancy by the drug, provided that the receptor-drug interactions obeyed the all-or-none rule at each available receptor site. Clark's theory was subsequently modified by KEIR (1971), among others, as it is now evident that all the receptor sites need not be occupied for a biological response to occur and, further, that receptor occupancy and the extent of a biological response are not directly correlated. These conclusions are germane to steroid-receptor interactions because there is now increasing evidence that an excess of steroid receptors is present in many mammalian cells (for review, see KING and MAINWARING, 1974). The principal stricture, however, to the use of "receptor" in the context of the mode of action of hormones is the widely reported observation that steroids can bind effectively to their intracellular receptors without promoting a hormonal response. This runs counter to the classical definition of a receptor in the pharmacologic idiom, namely the macromolecular component to which a drug must combine in order to elicit its biological response. The precise nature of the other intracellular elements involved in the manifestation of hormonal responses remains the topic of great speculation and interest. With these reservations, the term receptor features throughout the present text as it circumvents difficulties in otherwise describing the high-affinity binding of steroid hormones in intact cells and subcellular fractions.

2. The Model for the Mechanism of Action of Androgens

From studies undertaken in this laboratory and elsewhere, a model for the mechanism of action of androgens may be proposed (Fig. 1).

The interaction and retention of an androgen with its target cells is not a random process but is mediated by an integrated sequence of molecular events. (1) The principal androgen, testosterone, is transported in the plasma in the form of stable complexes with plasma proteins. (2) Testosterone enters all cells to a certain extent by a process that is currently not understood. (3) Exclusively within androgen target cells, testosterone is subjected to extensive metabolism and the principal metabolite is 5α-dihydrotestosterone (17β-hydroxy-5α-androstan-3-one). (4) 5α-Dihydrotestosterone is a potent androgen and it is specifically bound with a high affinity to a cytoplasmic receptor protein, forming an androgen-receptor protein complex. (5) By subtle alterations in the tertiary or quaternary structure, a change in the configuration of the 5α-dihydrotestosterone-receptor protein complex occurs, resulting in an activated complex with an increased propensity for the nuclear acceptor sites. (6) The activated androgen-receptor protein complex is translocated into the nucleus and is retained for a significant but finite period of time at fairly precisely defined sites in chromatin. (7) The advent of the androgen-receptor protein complex stimulates many biochemical events that mandatorily require DNA as template and thereby initiates the spectrum of processes responsible for the manifestation of the androgenic response. The hormone-mediated events proceed in an ordered temporal sequence but at markedly different rates. For simplicity, these will be classified as initial, early, and late events. (8) The androgen-receptor protein complex finally breaks down or is displaced from the nucleus by an as yet ill-defined mechanism. As a consequence, the critical biochemical processes implicated in the androgenic response slow down. To comply with the fundamental tenets of endocrinology, the entire system must be reversible; therefore, interactions between 5α-dihydrotestosterone and the cytoplasmic receptor protein and the subsequent association of the receptor protein-androgen complex with the nuclear acceptor components do not involve covalent bonds. In keeping with this premise, the androgenic steroid remains freely extractable into organic solvents throughout the entire entry and retention process.

As extensively discussed by KING and MAINWARING (1974), similar models can be proposed for the mechanism of action of other classes of steroid hormones, save that the metabolism of the naturally secreted hormone is a unique and characteristic feature of the androgens. The development of these experimental models is an outstanding success of contemporary endocrinology conducted at the molecular level and several factors have sustained this widespread interest and progress. First, the ubiquitous description of high-affinity receptor systems provided a novel and exciting

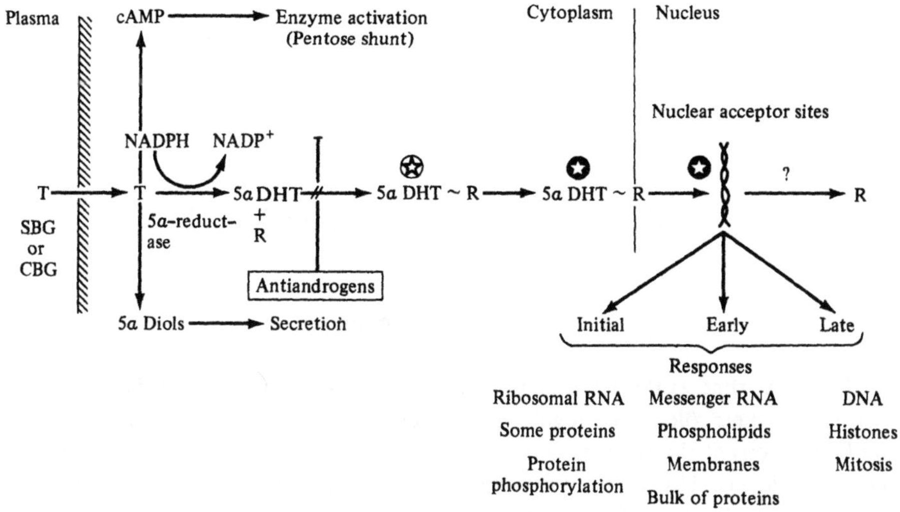

Fig. 1. A schematic model for the mechanism of action of androgens. *T* testosterone; *SBG* sex steroid-binding β-globulin; *CBG* corticosteroid-binding α_2 – globulin; *5αDHT* 5α-dihydrotestosterone; *5αDiols* 5α-androstanediols; ✪ and ✪ indicate changes in configuration of receptor complex during activation

insight into the regulation of biochemical pathways where only relatively small numbers of steroid molecules were involved. Second, studies on steroid receptor systems offered new hope in the diagnosis and treatment of clinical disorders with a hormonal aetiology. The importance of the measurement of oestrogen receptors in the successful management of human breast cancer is particularly germane in this context (FOLCA et al., 1961; MCGUIRE, 1973; SAVOLO et al., 1974; de SOMBRE et al., 1974). Third, experiments on the mode of action of steroid hormones followed in the wake of the impressive advances made in delineating the intracellular machinery for macromolecular syntheses, especially in bacteria and mammalian cells maintained in tissue culture. These technologic advances have been widely applied to hormone-sensitive cells, especially with respect to nucleic acid metabolism.

3. Experimental Support for the Model

a) Antiandrogens

Until such time as the receptor protein-hormone complexes are purified to homogeneity and their rôle in the regulation of biochemical reactions is precisely defined, the foundations of the receptor concept currently rest on impressive, but nevertheless indirect, evidence. However, by judicious use of antagonists of steroid hormones, the mandatory involvement of receptor proteins in the expression of the vast majority of hormonal responses is now established beyond reasonable doubt. Many compounds are able to suppress the biological activity of androgens and may thus be reasonably described as antiandrogens. The two most intensively studied antiandrogens, cyproterone acetate (6α-chloro-17α-acetoxy-1,2α-methylene-4, 6-pregnadien-3, 20-dione) and BOMT (6α-bromo-17α-methyl-17β-hydroxy-4-oxa-5α-androstan-3-one) are structurally related to steroids. Both counter the formation of receptor protein-5α-dihydrotestosterone complex (FANG and LIAO, 1969; MANGAN and MAINWARING, 1972; 1973), and negate most manifestations of the androgenic response. A wide spectrum of processes is impaired in androgen target cells by antiandrogens, including the induction of enzymes, such as acid phosphatase (GELLER et al., 1969) and type A aldolase (MAINWARING et al., 1974 b), ribosomal RNA synthesis, and nucleolar (form A) RNA polymerase activity (MAINWARING et al., 1971), polyribosome synthesis (MAINWARING and WILCE, 1973), poly(A)-rich messenger RNA synthesis (MAINWARING et al., 1974 c), polyamine synthesis (MANGAN et al., 1973) and many enzymes engaged in DNA synthesis (RENNIE et al., 1975). This comprehensive list indicates the very extensive involvement of the receptor system in the mechanism of action of androgens.

b) Experimental Testicular Feminisation Syndromes

Stable, genetically inherited defects in animals, similar to human syndromes of clinical importance, provide invaluable experimental systems for investigating the genetic regulation of sexual differentiation and metabolism. From such studies, the fundamental lesions responsible for many endocrinologic aberrations have been determined. Two such mutants provide experimental models for the testicular feminisation syndrome (male genotype; female phenotype), the pseudohermaphrodite rat (STANLEY and GUMBRECK, 1974) and the Tfm mouse (LYON and HAWKES, 1970). In both, androgenic control of sexual differentiation of the reproductive tract is lost, the external male genitalia fail to appear and the vestigial testes remain high in the inguinal canal, despite the presence of the X chromosome. The administration of even excessive doses of androgens fails to evoke any reversal of these profound abnormalities and the Tfm mutants are good examples of total end-organ insensitivity to hormones. Defects in the transport or metabolism of testosterone do not provide plausible explanations of these sex-linked disorders (BARDIN et al., 1970; BULLOCK et al., 1971 a; GROSSMAN et al., 1971). Furthermore, the synthesis of testosterone, although considerably impaired, is by no means completely repressed (BARDIN et al., 1969). Although the male accessory sexual glands are absent, studies on the binding of androgens can still be conducted on Tfm mutants because certain androgen-sensitive organs persist, including preputial gland, kidney, and submandibular gland. Despite certain references to the contrary (DREWS et al., 1972; WILSON and GOLDSTEIN, 1972), it would appear that the 7S cytoplasmic receptor and the nuclear binding mechanism, responsible for the specific binding of testosterone and 5α-dihydrotestosterone in normal mouse kidney (BULLOCK and BARDIN, 1972; BULLOCK et al., 1975), are totally absent in the Tfm mutants (BULLOCK and BARDIN, 1970; BULLOCK et al., 1971 b; GEHRING et al., 1971; RITZÉN et al., 1972; BULLOCK et al., 1975). Overall, a defective mechanism for the binding of androgens seems the most persuasive explanation of testicular feminisation and adds powerful weight to the receptor concept. Somewhat surprisingly, the work of DREWS et al. (1972) is diametrically opposed to this conclusion; they claimed that the nuclear binding of androgens was in no way impaired in the Tfm mouse. However, certain aspects of this study (DREWS et al., 1972) are not completely satisfactory. The binding studies were conducted at the remarkably low concentration of 2.2×10^{-12} M [^{3}H] 5α-dihydrotestosterone; this seems well below the physiologic range, at least in terms of the induction of kidney enzymes (TETTENBORN et al., 1972) and binding of this remarkably high affinity need not necessarily have biological relevance (see KING and MAINWARING, 1974). In addition, kidney cytoplasm can metabolise [^{3}H] 5α-dihydrotestosterone at a remarkable rate and such low concentrations of tracer steroid would be rapidly converted to a variety of products of unknown composition. Overall, studies on the Tfm mutants provide convincing evidence in support of the proposed model for the mechanism of action of androgens. These important studies are discussed in great detail in a companion monograph in this series by Dr. C. W. Bardin (Clinical and Experimental Disorders of Androgen Biosynthesis and Action).

c) Specificity of Induction and Dose-Response Phenomena

With presently available technology, credence in contemporary models of hormone action has also been based on the powerful criterion of the specificity of the hormonal responses. If models embodying the receptor concept are essentially correct, then the high-affinity binding of steroids must satisfy two fundamental requirements. First, maximum binding should occur at steroid concentrations within the physiologic range and second, the binding must accurately reflect the established biology of the hormonal response in terms of steroid- and tissue-specificity. For many classes of steroid hormones, such information is readily at hand. With the glucocorticoids, for example, the induction of tyrosine aminotransferase and a cell-adhesiveness factor in HTC cells precisely reflects the binding of steroids to the glucocorticoid receptor system (SAMUELS and TOMKINS, 1970; BAXTER and TOMKINS, 1971; ROUSSEAU et al., 1972). An equally rigid association between the binding of steroids and their relative influence on glucose uptake in thymus-derived lymphocytes has also been presented (MUNCK and BRINCK-JOHNSEN, 1968; MUNCK and WIRA, 1971). For many reasons, comparable studies on androgen-responsive cells are remarkably few. Citing glucocorticoid-sensitive systems for the basis of comparison, studies may be performed with the advantages of the cell culture technique *in vitro* and the hormonal response is manifest either in the induction of a well-characterised protein or in a specific biochemical process. These features are not so evident with androgen-sensitive systems; the authentic stimulation of biochemical events under conditions *in vitro* has not been widely documented and the induction of specific protein markers has not been extensively reported. The induction of prostate aldolase (MAINWARING et al., 1974 b) does not readily lend itself to accurate assessment of dose-response phenomena because the androgenic stimulus necessarily had to be administered to animals *in vivo*.

Recent studies on the induction of foetal haemoglobins (types E and P) in organ cultures of chick blastoderm by 5β-reduced metabolites of testosterone *in vitro* do fulfil the specificity embodied in the general model of hormone action (IRVING et al., 1976). Strictly speaking, this process is not mediated by androgenic steroids but by the 5β-reduced steroids that may be formed in the blastoderm from testosterone and other precursors, such as cholesterol, at a time commensurate with the onset of erythropoiesis (PARSONS, 1970). 5β-Reduced steroids have no classical androgenic activity in conventional bioassay procedures based on the weight regain of male accessory sexual glands (see LIAO and FANG, 1969). In addition, the induction process, although primed by the metabolism of testosterone, is not linked to sexual differentiation. Changes in the induction of foetal haemoglobins evoked by aetiocholanolone are evident only during the early period of embryonic development, some 24–60 h after fertilisation; sexual differentiation occurs at a much later developmental stage (ROMANOV, 1960). Enhancement of the synthesis of foetal haemoglobins occurs at very low concentrations of aetiocholanolone in the physiologic range 10^{-10} to 10^{-8} M and a strict dose-response relationship was established (IRVING et al., 1976). Furthermore, from the evidenced summarised in Table 1, the steroid specificity of the induction process is clearly reflected in the relative binding of steroids in soluble extracts of chick blastoderm. The contrasting influence of the planar (*trans* A:B rings) 5α- and extremely angular (*cis* A:B rings) 5β-stereoisomers of dihydrotestosterone is particularly impressive. From recent experience, the 5β-specific receptors in chick blastoderm are the most labile of the high-affinity binding components studied in this laboratory over recent years and further characterisaton of these receptors has not yet been accomplished. The existence of 5β-specific receptors has recently been confirmed by LANE et al. (1975) in cultures of foetal chick hepatocytes. Differences in binding of the 5α- and 5β-stereoisomers of dihydrotestosterone were less dramatic than in blastoderm but this binding mechanism may well be implicated in the regulation of haem biosynthesis in these foetal liver cultures. Taken in concert, these results strongly corroborate the principal features of the model for the mechanisms of action of steroids and their metabolites. A given system selectively forms the metabolites of testosterone most suited to the biological response and metabolic regulation is initiated by a specific receptor mechanism.

Table 1. The relative binding of [³H]-labelled steroids in cytoplasmic extracts of chick blastoderm and their ability to induce foetal haemoglobin.

After 24 h of embryonic development, chick blastoderms were dissected from fertilised eggs and maintained as organ cultures, in vitro. For binding experiments, cytoplasmic extracts were prepared from 36 pooled blastoderms after 12 h in culture and [³H]-labelled steroids were added (10,000 c.p.m./0.25 ml of extract). After incubation for 2 h at 0° C, specific or high affinity binding was determined by DNA-cellulose chromatography (MAINWARING and IRVING, 1973). For induction experiments, steroids were added to concentration of 30 nM and after 48 h of culture, foetal haemoglobins E and P were determined in tissue pooled from six blastoderms. Results are taken from IRVING et al. (1976) Biochem. J. **154**, 83 and unpublished work from this laboratory.

Steroid	Induction of haemoglobin (μg/20μg of blastoderm DNA)	Binding of ³H steroid (c.p.m./500 mg. wet wt. of DNA-cellulose)
None	10.5	-
5β-Dihydrotestosterone	37.5	760
Aetiocholanolone	38.5	N.D.
5α-Dihydrotestosterone	11.5	280
Cortisol	15.4	220
Oestradiol-17β	7.1	190
Progesterone	23.0	N.D.
Testosterone	N.D.[a]	180[b]

[a] *N.D.* = not determined; [b] specific activity of all [³H] steroids was 11–13 Ci/mmole.

d) Temporal Integration of Steroid-Mediated Responses

In the model (Fig. 1), it is envisaged that the expression of a hormonal response proceeds by an integrated series of steps and with an ordered temporal sequence. As a corollary, the enhancement of early events set in train by the advent of the hormone subsequently initiates processes occurring later in the overall biological response. In general terms, evidence implicating a receptor mechanism in the crucial initial events is overwhelming. In later chapters, other examples of the early biochemical events promoted by androgens will be discussed, but none of them has the necessary specificity to denigrate the mandatory involvement of the receptor system in the early, if not the earliest, interactions between androgens and their target cells. The importance of studies on antiandrogens in this context has already been emphasized earlier in section I.3.a.

Corroboration of the integrated sequence of temporal events is provided by a wealth of studies with metabolic inhibitors, especially actinomycin D, cycloheximide, and puromycin. Suppression of the early synthesis of RNA and protein by the concomitant administration of inhibitors and testosterone *in vivo* prevents enzyme induction, growth, and mitosis. As an example of such studies, the investigation by Coffey and his associates on the regulation of DNA synthesis by androgens is cited (CARTER et al., 1972; SUFRIN and COFFEY, 1973). When testosterone is administered daily to male rats 7 days after bilateral orchidectomy, there is a dramatic enhancement of DNA replication, expressed either as mitotic indices or replicative DNA polymerase activity, but only after a protracted delay or latent period of 48–72 h. The surge in these indicators of cell proliferation could be prevented either by inhibitors of protein synthesis or antiandrogens (SUFRIN and COFFEY, 1973). The degree of inhibition elicited by cycloheximide and puromycin was in accord with their suppression of nuclear protein synthesis (CHUNG and COFFEY, 1971) and the effects of antiandrogens were directly related to their competition for the binding sites within the androgen receptor system (CARTER et al., 1972; SUFRIN and COFFEY, 1973; MAINWARING et al., 1974 a). Of prime importance, however, metabolic inhibitors or antiandrogens were active only during the first 24 h of the latent period when the early phases of the mitotic response were set in motion; at later times they were totally ineffective (CARTER et al., 1972). These astutely designed experiments unequivocally indicate an ordered temporal sequence of biochemical events in the instigation of androgenic responses.

e) Tumours with Different Sensitivities to Androgens

Experimental androgen-responsive tumours are not common, in contrast to the situation existing with oestrogen- and particularly glucocorticoid-sensitive tumours. From a spontaneous mammary tumour in a primipara female mouse, MINESETA and YAMAGUCHI (1965) developed two tumour sublines, the androgen-dependent Shionogi 115 and the androgen-independent Shionogi 42 tumours. There is no morphologic or karyotypic evidence to suggest that these tumours originated by the mutation or reversion of a common malignant cell, but despite this reservation, the S115 and S42 tumours do provide accessible systems in which the influence of androgens on the growth of tumours can be ascertained. Both the S115 (YAMAGUCHI et al., 1971; MAINWARING and MANGAN, 1973) and the S42 (MAINWARING and MANGAN, 1973) tumours contain the NADPH-dependent 5α-reductase enzyme necessary for forming 5α-dihydrotestosterone, although the activity is somewhat higher in S115 cells (MAINWARING and MANGAN, 1973). The relative binding of androgens in these two tumours has provided a more convincing explanation of their contrasting responses in terms of growth to hormonal manipulation. MATSUMOTO et al. (1972) demonstrated a higher uptake of [^{3}H]-labeled androgens in the S115 tumour than in the S42 tumour, but the intracellular distribution of the radioactive hormones was not rigorously examined. In a more comprehensive study, BRUCHOVSKY (1972) reported that after the injection of [^{3}H]-testosterone into tumour-bearing mice, small amounts of [^{3}H]-5α-dihydrotestosterone were associated with tumour nuclei but significantly less than in the accessory sexual glands, such as seminal vesicle. An interesting facet of this study by BRUCHOVSKY (1972) was that the S115 tumour displayed a reduced preference for the binding of 5α-dihydrotestosterone than the accessory sexual glands and testosterone constituted about half of the androgens retained in the tumour nuclei. MAINWARING and MANGAN (1973) extended this comparison of androgen binding and established a higher nuclear binding of androgens in the S115 as compared with the S42 tumour; as BRUCHOVSKY (1972) found, however, the nuclear binding mechanism even in the S115 tumour was less efficient than that in the male accessory sexual glands of the tumour-bearing mice. At this point, it appeared that differences in the binding of androgens could explain the different behaviour of the S115 and S42 tumours to androgenic stimulation.

Based on the earlier report that the S115 tumour remained responsive to androgens after prolonged maintenance under conditions of cell culture (MEAKIN and ROBINS, 1968), further interest in this tumour was aroused (SMITH and KING, 1972; SUTHERLAND et al., 1974; RENNIE and BRUCHOVSKY, 1972). In a considerable advance over previous investigations, GORDON et al. (1974) demonstrated the presence of both cytoplasmic and nuclear androgen receptor proteins in cultured S115 cells and confirmed the significant high-affinity binding of testosterone (BRUCHOVSKY, 1972). However, as more effort was concentrated on the S115 tumour in culture, one disquieting feature of its behaviour became generally acknowledged, namely that the tumour cells grew reasonably well even in the total absence of testosterone and that androgens only enhanced the rate of growth in chemically defined media *in vitro* (SUTHERLAND et al., 1974; GORDON et al., 1974). For this reason, the interpretation of the earlier studies on the relative binding of androgens in the S115 and S42 is in doubt. The Shionogi tumour system is clearly more complicated than was first thought. The spontaneous prostate tumours in Lobund rats described recently by POLLARD and LUCKERT (1975) may feature prominently in future investigations on the regulation of tumour growth by androgens.

4. Limitations of the Model of Androgen Action

a) Fundamental Mechanisms of Hormone Action

As discussed at length by KING and MAINWARING (1974), models for the mechanism of action of all steroid hormones are remarkably similar and despite profound differences in the nature of the biological responses, all are believed to be initiated by receptor protein-hormone interactions. All responses essentially fall into two distinct categories, switch (inaugural) processes or amplification (modulation) phenomena. In fundamental terms, switch processes may be equated with the qualitative changes promoted by steroid hormones whereas amplification processes embrace purely quantitative changes. Switch processes controlled by androgens are surprisingly few, but based primarily on the classical studies of JOST (1953; 1967; 1970) it is now accepted that the development of the male phenotype and the differentiation of the male urogenital tract, except for the regression of the Müllerian duct, is promoted by the secretion of testosterone from the foetal testis. This profound change during a critical phase of embryonic development is a prime example of a switch mechanism controlled by androgens. Nearly all the other processes regulated by androgens, especially in adult animals, may safely be described as amplification phenomena. Inspection of Figure 1 fails to show how receptor mechanisms alone can initiate switch as against amplification processes. Other components in the genetic apparatus must be implicated in establishing the mode of hormonal responses, but the nature of these subtle regulators is totally unknown.

b) Redundant (Aberrant) Binding

The current literature contains examples of the anomalous situation where the high-affinity binding of steroids is not commensurate with a biological response. Taking but two examples, there is no response to the binding of oestrogens in animals with obligatory delayed implantation (O'FARELL and DANIEL, 1971) and certain sublines of transitorily sensitive lymphoma cells (for example, S1 AT.8) contain a constant complement of specific receptor proteins despite fluctuating sensitivities to the cytolytic effects of glucocorticoids (BAXTER et al., 1971). Androgen target cells provide few examples of such phenomena. Contrary to the consensus of research opinion presented in section I.3.b, very extensive binding of testosterone has been observed to an atypical 3S binding protein in the submandibular gland of the Tfm mouse (WILSON and GOLDSTEIN, 1972) yet development of the male genitalia is totally suppressed. High-affinity binding of [^{3}H]-5α-dihydrotestosterone in adult rat testis has been demonstrated (MAINWARING and MANGAN, 1973) yet the 5α-reductase necessary for forming this metabolite of testosterone is absent. By developmental restriction of steroid-metabolising enzymes, the activity of certain androgen receptor systems can be severely curtailed. From these examples, it is clear that receptor proteins can be demonstrated experimentally in target cells at times when the hormonal response is suppressed by regulatory factors or components yet unknown.

c) General Applicability of the Model

The model is applicable to the mode of action of androgens in the accessory sexual glands of most experimental animals, but this generalisation must be accepted with a certain measure of caution. The secondary sexual glands of mammalian species display discrete species-specific differences, a good example being the prostate of the dog. The dog alone shares with man a disposition to abnormalities in prostatic growth during old age, resulting in benign prostatic hyperplasia and metastasizing prostatic carcinoma. In an exciting new development, Pierrepoint and his collaborators (HARPER et al., 1971; EVANS and PIERREPOINT, 1975) now have irrefutable evidence that the active androgen in the dog is not 5α-dihydrotestosterone, as in man (PIKE et al., 1970; MAINWARING and MILROY, 1973), but 5α-androstan-3α, 17α-diol. This work raises several interesting questions. First, even among the male accessory sexual glands, models of androgen action based exclusively on the formation and binding of 5α-dihydrotestosterone may not be generally applicable. Second, it raises serious limitations in the adoption of the canine prostate as an experimental model for human prostatic carcinoma, for aside from anatomic or physiologic differences, the two glands are probably subject to different androgenic stimuli. These considerations are also important to the widespread screening of antiandrogens of putative chemotherapeutic value in the dog.

The anabolic function of androgens in muscle, expressed usually as the maintenance of a positive nitrogen balance, cannot be explained by the simplistic model. With perhaps one notable exception, the levator ani or bulbocavernosus muscle of guinea pig (MAINWARING and MANGAN, 1973), skeletal muscle has an exceedingly limited ability to form 5α-dihydrotestosterone (WILSON and GLOYNA, 1970; KELCH et al., 1971; MAINWARING and MANGAN, 1973). Furthermore, as discussed later in Chapter III, the high-affinity binding of androgens in muscle is currently the centre of contentious debate.

d) Acute Tissue-Specificity of Androgenic Responses

The stringent tissue specificity of the responses evoked by androgens is truly remarkable, even among androgen-sensitive organs in anatomical juxtaposition and hence receiving a similar if not identical degree of androgenic stimulation. For example, certain accessory sexual secretions characteristically contain fructose while others exclusively contain citric acid (SAMUELS et al., 1962), alcohol dehydrogenase and specific isoenzymes of β-glucuronidase are inducible only in mouse kidney (OHNO et al., 1970; SWANK et al., 1973) and certain basic proteins are a unique feature of the secretion of rat seminal vesicle (MÁNYAI et al., 1965; TÓTH and ZAKÁR, 1971). The experimental model cannot explain these subtle differences in androgenic responses. During the course of differentiation, the structure of chromatin must be modified in such a manner that the association of the androgen-receptor protein complex triggers a truly unique and tissue-specific response. Analysis of such sophisticated control mechanisms is beyond the limits of available technology, but since the genetics of the induction process in mouse kidney are advanced and the structure of β-glucuronidase is now well established (SWANK and PAIGEN, 1973), the mouse kidney system is a likely candidate for studies on acute tissue specificity in the future. From recent experience in this laboratory, the rat seminal vesicle system also has much to offer in elucidating the molecular basis for the stringent tissue specificity of androgen action. Two basic glycoproteins are induced exclusively in seminal vesicle by androgens and the receptor mechanism is undoubtedly involved. Since these proteins may be readily purified and together constitute 30–35% of the total protein in seminal secretions (HIGGINS et al., 1976), it should be possible to isolate the messenger RNA coding for these proteins and to establish whether its synthesis is regulated in a uniquely tissue-specific manner.

e) Androgen Responses Independent of the Receptor System

Androgenic responses independent of the high-affinity binding mechanism are not well documented but several have been described. A great success of contemporary biochemistry in terms of metabolic regulation has been the development of the second messenger concept by Sutherland and his collaborators. This depends on the intracellular synthesis of adenosine-3′, 5′-cyclic monophosphate (cyclic AMP) after hormonal stimulation of the enzyme, adenyl cyclase. SINGHAL et al. (1971) demonstrated an activation of prostate adenyl cyclase by testosterone, together with the induction of many enzyme activities and claimed that the second messenger concept was of outstanding importance in the mechanism of action of androgens. MANGAN et al. (1973) challenged this viewpoint, with independent support being provided later by CRAVEN et al. (1974). While it remains true that cyclic AMP can modulate the activities of certain enzymes (MANGAN et al., 1973), these are extremely restricted in number and essentially limited to the pentose phosphate cycle. Of extreme interest was the observation that the processes stimulated by cyclic AMP were totally refractory to the antiandrogen, cyproterone acetate (MANGAN et al., 1973), indicating that they were not mediated by the androgen receptor system. In later studies, it was found that cyclic AMP could not mimic testosterone in regulating the rate of growth and cell division of the rat prostate gland (CRAVEN et al., 1974). Another example of a receptor-independent process is given later in section I.4.g.

f) Hormonal Synergism

Prolactin can directly enhance the uptake of testosterone into the accessory sexual glands of the rat (GRAYHACK and LEBOVITZ, 1967) and the dog (RESNICK et al., 1974) and conversely, hypophysectomy reduces the uptake of testosterone into the rat prostate (LAWRENCE and LANDAU, 1965). These observations on hormonal synergism are difficult to reconcile with the present model of androgen action because the influence of the pituitary hormone could be exerted at many stages of the overall sequence of reactions. Furthermore, the biological significance of these studies is still obscure but the remarkable observation that the effects of prolactin were specific to certain accessory sexual glands only, even in the same animal (GRAYHACK and LEBOVITZ, 1967), raises the interesting possibility that pituitary hormones may regulate the distribution and hence the effects of androgens in a very subtle way. To answer this question unequivocally, more research on hormonal synergism is needed and further results are awaited with interest.

g) Responses Independent of Transcription

Androgens can exert a profound influence on transcriptional events, as indicated by the report from this laboratory on the synthesis of the messenger RNA for prostate (type A) adolase (MAINWARING et al., 1974 b). However, it would be erroneous to surmise that all androgenic responses are mediated exclusively by nuclear processes and, by inference, an enhancement of genetic transcription.

Certain evidence suggests that the induction of β-glucuronidase activity may proceed exclusively in the cytoplasmic compartment of mouse kidney by purely translational mechanisms. Comparison of enzyme induction and kidney hypertrophy after androgenic stimulation, together with the effects of antiandrogens, have provided some interesting observations. 5α-Dihydrotestosterone promoted a prolonged period of enzyme induction and hypertrophy, whereas 5α-androstanediols caused instant but transient enzyme induction but no hypertrophy (OHNO et al., 1971). Both kidney hypertrophy and the nuclear binding of androgens, but not enzyme induction, were inhibited by the concurrent administration of cyproterone acetate with testosterone *in vivo* (OHNO and LYON, 1970). The nuclear binding of androgens in mouse kidney under conditions *in vivo* (BULLOCK et al., 1971) and *in vitro* (MAINWARING and MANGAN, 1973) is exclusive for testosterone or 5α-dihydrotestosterone; nuclear binding of 5α-androstanediols has not been reported and these metabolites of testosterone are predominantly bound in the cytoplasmic microsomal fraction (BAULIEU et al., 1971; ROBEL et al., 1974). Studies by SWANK and PAIGEN (1973) indicate that the androgen-inducible or X form of β-glucuronidase is specifically located in the microsomal fraction of mouse kidney and taken overall, these results support a cytoplasmic or translational mode of androgenic regulation for this enzyme by 5α-androstanediols.

Similar considerations possibly apply to the amplification of foetal haemoglobin synthesis in cultures of chick blastoderm by aetiocholanolone. The messenger RNA coding specifically for the polypeptide chains of foetal haemoglobins E and P has been isolated and translated with fidelity in messenger RNA-depleted system for protein synthesis derived from wheat germ (IRVING et al., 1976). More detailed enquiries revealed that the globin messenger RNA was present throughout the early stages of development of chick blastoderm, even in the presence of the 5β-steroid inducer, but was not translated. These 5β-reduced steroids certainly enhance the synthesis of haemoglobins E and P but only at a later developmental stage. The most plausible explanation of these results is that aetiocholanolone and other angular steroids, with their A and B rings characteristically in the *cis* orientation, can regulate the synthesis of foetal haemoglobins by translational rather than transcriptional control mechanisms.

As a final example, SUFRIN and COFFEY (1973) reported that the stimulation of DNA synthesis in rat prostate gland by androgens was not sensitive to the metabolic inhibitor, actinomycin D, particularly during the latent period of this late androgenic response. Again, the most plausible explanation of these findings is that the enhancement of the activities of certain enzymes associated with DNA replication does not require de novo RNA synthesis. Androgens could elicit this effect either by modification of preexisting enzyme moieties or by stimulating the translation of stable messenger RNA already present before androgenic stimulation.

5. Conclusions

A plausible mechanism of action for androgens has been proposed on the basis of experimental findings reported in the current literature. It should be stressed that the model is largely based on studies conducted on the rat ventral prostate gland and there are serious reservations in applying the model to all androgen target systems, even to other accessory sexual glands. For the main part, these reservations are expected, because androgens can regulate a bewildering spectrum of biological responses in a diversity of cell types.

In the chapters that follow, the metabolic processes under androgenic control will be discussed in greater detail and an attempt will be made to interpret present knowledge in the light of the known biology of different androgen target cells.

II. The Metabolism of Androgens in Relation to Their Mechanism of Action

Progestational steroids may be metabolised within certain organs as a necessary and integral step in their mode of action; for example, 5α-pregnan-3,20-dione is formed in the shell gland of the domestic fowl (MORGAN and WILSON, 1970) and various 5α-reduced metabolites of progesterone appear in significant amounts in rat uterus (WIEST, 1970; ARMSTRONG, 1970; REEL et al., 1971). With these notable exceptions, the enzymic transformation of a naturally secreted hormone within its target cells is a characteristic feature of the androgens. After a rather confusing period of research on androgen metabolism and transport in the early 1960's (for review, see KING and MAINWARING, 1974), the first clear indication that 5α-dihydrotestosterone could be formed within accessory sexual glands was obtained by FARNSWORTH and BROWN (1963) from arterial infusion of the ventral lobe of rat prostate with [^{3}H] testosterone. The important metabolite, 5α-dihydrotestosterone, was identified by paper chromatographic analysis of the efferent blood from the prostatic vein. This innovative work was later substantiated by ANDERSON and LIAO (1968) and BRUCHOVSKY and WILSON (1968), with the further important observation that the 5α-dihydrotestosterone was firmly associated with the nuclei of male accessory sexual glands. The resurgence of interest in the mechanism of action of androgens may be attributed to the impact of these enterprising studies.

Admirable reviews of early work on the anabolism and catabolism of androgens are available (DORFMAN and SHIPLEY, 1956; DORFMAN and UNGAR, 1965; OFNER, 1969); recent investigations are the subject of a companion monograph in this series by Dr. T. Tamaoki (Androgens; their Biosynthesis and Related Biochemistry). However, the mechanism of action of androgens cannot be considered in depth without extensive reference to androgen metabolism. In this chapter, a résumé of androgen metabolism is presented so that the interpretation of contemporary studies on the mode of action of androgens will be facilitated.

A particularly striking aspect of androgen metabolism is its dramatic variation from one androgen target cell to another and interspecies differences are the rule rather than the exception. This adds a novel degree of complexity to the mechanism of action of androgens. Phylogenetic studies on androgen metabolism are not sufficiently advanced to establish when this characteristic attribute of androgen target cells was acquired during the evolutionary process. Nevertheless, androgen metabolism is probably an invariant feature of the Chordata and must be considered an evolutionary advantage because it provides a subtle mechanism whereby a single hormonal stimulus, the parent androgen, may be modulated to elicit a varied spectrum of responses. From this, it follows that experiments on the mechanism of action of androgens must be designed with care and it is imperative that a critical appraisal of the metabolic potential of a given experimental system must be made before other biochemical studies are undertaken. This caveat is particularly relevant to studies on the binding of androgens. To illustrate the significance of intracellular metabolism, the submaxillary salivary gland of the boar forms appreciable quantities of 5α-dihydrotestosterone (BOOTH, 1972) yet the corresponding gland of the dog transforms testosterone almost exclusively to androstenedione rather than 5α-reduced metabolites (WEINER et al., 1970). Thus, binding of 5α-dihydrotestosterone is likely to be of biological importance in the submaxillary gland of the boar but not the dog. Strictly in accordance with the original definition by STARLING (1905), 5α-dihydrotestosterone and other biologically active metabolites of testosterone should not be termed androgens. To overcome this difficulty, they are generally described as active metabolites or even active androgens in the current literature; the latter description, however, is a contradiction in terms and active metabolites is preferred in the present text. This choice is not made on pedantic grounds, because one of the few functions of testosterone that cannot be mimicked by its metabolites is the maintenance of male sexual behaviour (MCDONALD et al., 1970; WHALEN and LUTTGE, 1971 a); this is the classically defined function of an androgen.

1. Relative Biological Activities of Testosterone and Its Metabolites

From 1940 onward, a remarkable array of naturally occurring steroids was isolated and the synthesis of complex steroids also became a major facet of the pharmaceutical industry. Literally thousands of compounds must have been tested for androgen-like activity to date in a variety of bioassay procedures. These have ranged from an increase in size of the comb of the capon after topical application, increases in the weight of the accessory sexual glands, or levator ani (bulbocavernosus) muscles in immature animals after systemic or subcutaneous injection to complex assays based on patterns of sexual behavior. From this wealth of literature, the comparative biological activities of many naturally occurring steroids are cited in Table 2 from the reference treatise by DORFMAN and SHIPLEY (1956). Essentially consistent findings were reported by HUGGINS and MAINZER (1957) and SAUNDERS (1963) on the basis of other screening tests.

From such comparisons, it is evident that the biological activities of the metabolites of testosterone vary enormously; the androgenicity of certain metabolites can exceed that of testosterone whereas others have minimal biological potency. In certain bioassay procedures, 5α-dihydrotestosterone is more active than testosterone whereas the 5α-androstanediols generally have lower biological activity.

Inspection of the structures of these steroids (Fig. 2) provides some insight into the substituent groups necessary for maximal biological activity. Unquestionably, a 17-hydroxy group in the β-plane is a mandatory requirement for androgenicity, because testosterone and 5α-dihydrotestosterone have powerful androgenic properties, whereas androstanedione and 5α-androstan-3-one do not (LIAO et al., 1973 a). An oxygen function at the C-3 position is also necessary for biological activity as 17β-hydroxy-5α-androstane, for example, is inactive. As discussed in greater detail in Chapter III.1.c, the directive influence of the C-3 ketone group is particularly important in controlling the compartmentation of testosterone metabolites within androgen target cells. Metabolites containing the C-3 ketone group are actively retained within the nucleus, whereas 5α-androstanediols, with a hydroxy group at the C-3 locus are retained predominantly in the cytoplasmic microsomes (ROBEL et al., 1974). Oxygenation of certain steroids at the C-3 and C-17 positions can occur at sites outside androgen target cells, including liver (WOLFF and KASUYA, 1972), thus permitting certain compounds to display limited androgenicity, but only in biological trials of long duration. The presence of the angular methyl groups is not obligatory for the expression of androgenic activity, as 19-nortestosterone derivatives are active in all androgen target cells with the exception of skeletal muscle and brain. The insertion of additional methyl groups, especially in the 7α-position, also enhances

Table 2. Comparative activities of androgens on the seminal vesicles of immature male rats and the comb of the capon.

Data are taken from DORFMAN and SHIPLEY (1956) and reproduced with permission of John Wiley and Sons (New York)
Reference: testosterone = 100

	Comparative activities	
	Seminal vesicles	Comb
Testosterone	100	100
5α-Dihydrotestosterone (17β-hydroxy-5α-androstan-3-one)	200	75
17-Methyl-3α, 17β-dihydroxy-5α-androstane	50	50
3α, 17β-Androstanediol (3α, 17β-dihydroxy-5α-androstane)	33	75
Androstenedione (4-androsten-3, 17-dione)	20	12
Androstanedione (5α-androstan-3, 17-dione)	14	12
3β, 17β-Dihydroxy-5-androstene	14	3
3β, 17β-Androstanediol (3β, 17β-dihydroxy-5α-androstane)	10	2
Androsterone (3α-hydroxy-5α-androstan-17-one)	10	10
5-Androsten-3, 17-dione	7	12
Dehydroepiandrosterone (3β-hydroxy-5-androsten-17-one)	3	16
Epiandrosterone (3β-hydroxy-5α-androstan-17-one)	3	2

biological activity dramatically. Sophisticated molecular models (LIAO et al., 1973 a) and x-ray crystallographic analysis (COOPER et al., 1968) suggest that active metabolites of testosterone are relatively planar structures, whereas in testosterone, the A ring is considerably strained and distorted geometrically towards the α-face of the steroid ring system. This may well explain why testosterone has a lower binding affinity for many androgen receptor proteins (LIAO et al., 1973 a) and substituents reducing the distortion in the A ring or promoting a more planar structure overall will enhance biological activity. For such reasons, the synthetic steroid, 7α,17α-dimethyl-17β-hydroxy-4-estren-3-one (7α,17α-dimethyl-19-nortestosterone) has a biological potency exceeding that of testosterone and steroids containing conjugated double bonds from rings A to B to C, such as 17α-methyl-17β-hydroxy-estra-4,9,11-triene-3-one, also promote the growth of male accessory sexual glands (LIAO et al., 1973 a). Quite surprisingly, certain unusual compounds have some androgenic activity, including 1,4-*seco*,2,3-*bis*nor-17β-hydroxy-5α-androstane (ZANATI and WOLFF, 1973). The distinctive feature of these compounds is the absence of a formal A ring structure, again indicating perhaps the mandatory importance of the structure of the D ring, especially at the C-17 locus. X-ray crystallographic analysis of these compounds would be extremely important, because it would give a clear indication of the fundamental structural requirements for androgen-like activity.

In general terms, other classes of steroid hormones have negligible androgen activity. Progesterone was reported earlier to have limited influence on male sexual characteristics (GREENE et al., 1940) but this may be explained either by sample contamination with androgens or peripheral conversion to testosterone.

Thus far, the structure and activity relationships have been made with reference to the androgenic potency demonstrable in classical bioassay procedures. Without exception, these active compounds either have a Δ^4-double bond in the A ring (for

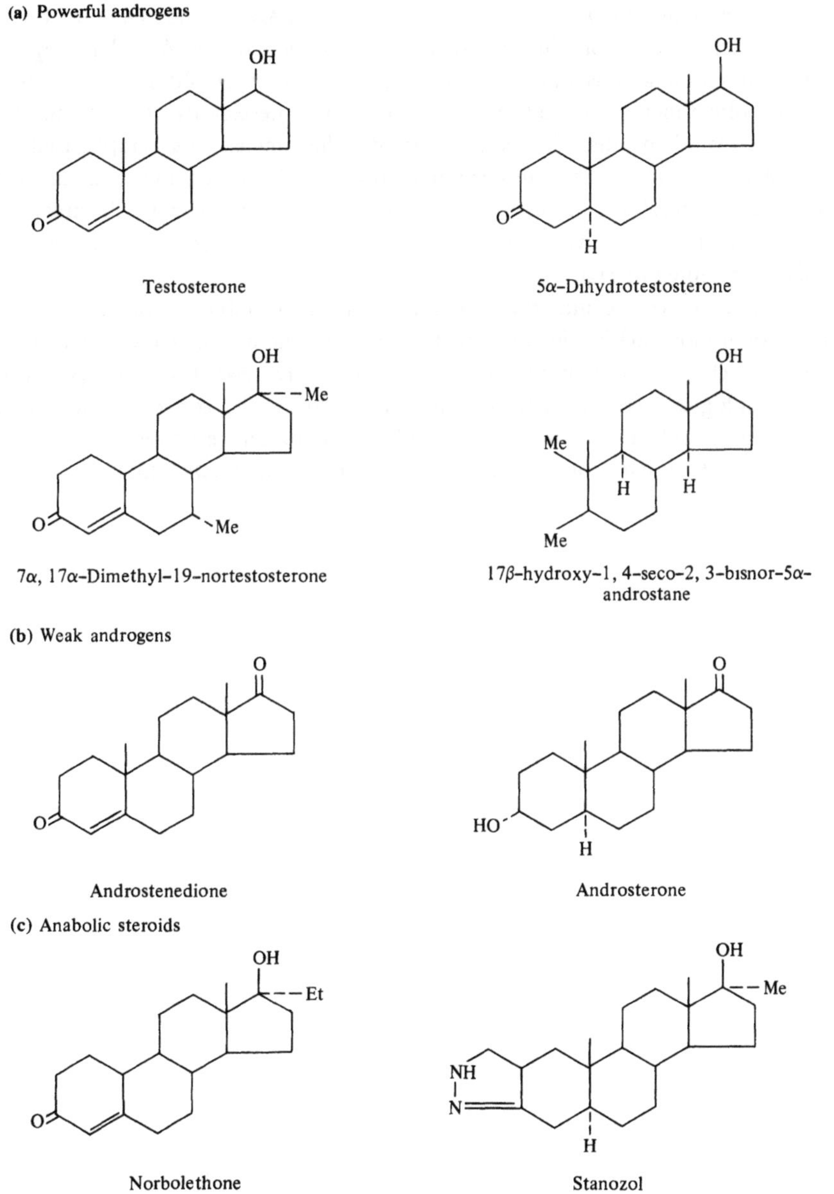

Fig. 2. Structures of powerful androgens, weak androgens, and anabolic steroids. In all structures, *Me* methyl and *Et* ethyl

example, testosterone) or are 5α-reduced metabolites (for example, 5α-dihydrotestosterone). In both categories, the A and B rings are in the relatively planar or *trans* configuration. By contrast, 5β-reduced metabolites of testosterone, such as aetiocholanolone, have an extremely angular structure resulting from the *cis* orientation of the A and B rings. From the classical investigations of Granick and his collaborators, 5β-reduced metabolites are specific inducers of haem synthesis in many experimental systems (LEVERE and GRANICK, 1965; GRANICK and KAPPAS, 1967). These important observations indicate that the orientation of the A and B rings results in a marked dichotomy in the biological activities of metabolites of testosterone.

The anabolic function of androgens also need not necessarily involve the substituent groups demanded by sex-associated phenomena. As emphasized by KOCHAKIAN (1975) in an authoritative review of androgenic and anabolic steroids, the structural requirements are quite different. In order to illustrate this point, the structures of two very powerful anabolic steroids, norbolethone and stanazol are presented in Figure 2.

From this succinct account, the importance of the metabolism of androgens is very clear. By minor modifications in structure, testosterone may be converted to active metabolites such as 5α-dihydrotestosterone or to closely related steroids possessing negligible biological activity such as androsterone. Metabolic conversion in the accessory sexual glands potentiates and modulates biological responses, whereas the catabolic processes in liver counter excessive androgenisation (DORFMAN and UNGAR, 1965).

2. General Comments on the Metabolism of Androgens

In recent years, the metabolism of androgens has been intensively studied and the majority of the enzymic steps are now well documented. Particular attention has been devoted to the NADPH-dependent Δ^4-3-ketosteroid-5α-reductase (more simply, 5α-reductase) responsible for the formation of the critical metabolite of testosterone, 5α-dihydrotestosterone. This enzyme utilises reducing equivalents of NADPH in a quite unique way; rather than being an indirect hydrogen donor as in steroid hydroxylation or hydrogenation processes, NADPH is used for the direct reduction of C=C double bonds. The 5α-reductase in male secondary sexual glands is principally located in the microsomal fraction (CHAMBERLAIN et al., 1966; MORFIN et al., 1970) but an appreciable amount is also associated with nuclear fraction (BRUCHOVSKY and WILSON, 1968) and more specifically, the nuclear membrane (MOORE and WILSON, 1972). This dual localisation is not surprising as the dynamic turnover of membranes in the prostate is regulated by androgens (MAINWARING and WILCE, 1972) and the processing of membranes of the endoplasmic reticulum is considered to occur on the outer nuclear membrane (MIRSKY and OSAWA, 1961). On kinetic and other criteria, the nuclear and cytoplasmic forms of 5α-reductase in the prostate are functionally and structurally related (FREDERIKSEN and WILSON, 1971). From studies on steroid substrate specificity (MCGUIRE and TOMKINS, 1960) and induction phenomena by medroxyprogesterone (GORDON et al., 1971; FREDERIKSEN and WILSON, 1971), however, prostate and liver 5α-reductases are quite distinct. The latter enzyme is implicated solely in steroid catabolism. In a comprehensive survey, WILSON and GLOYNA (1970) found a significant amount of 5α-reductase in nearly all male accessory sexual glands and enzymic activity was approximately correlated with the overall growth of the prostate in a given species. For the main part, the 5α-reductase activity is more than adequate for the synthesis of 5α-dihydrotestosterone and the reaction is limited solely by the intracellular concentrations of available testosterone (FREDERIKSEN and WILSON, 1971).

Despite its critical importance, 5α-reductase is not a ubiquitous feature of androgen target cells. There is unanimity among most investigators that the enzyme is not present in the skeletal muscles of most species (WILSON and GLOYNA, 1970; KELCH et al., 1971; MAINWARING and MANGAN, 1973), including the masseter and bulbocavernosus muscles used as the biological reference in screening tests for anabolic steroids. Exceptions to this general premise are few, but include the bulbocavernosus muscle of guinea pig (MAINWARING and MANGAN, 1973) and rabbit skeletal muscle (THOMAS, 1968). The importance of this latter study is somewhat clouded by the fact that enzymic activity was not measured quantitatively or rigorously compared with the 5α-reductase present in the accessory sexual glands. It appears that the formation and binding of 5α-dihydrotestosterone cannot play a significant rôle in the anabolic effects elicited by androgens in muscle.

In certain species, the activity of 5α-reductase is influenced by the circulating concentrations of androgens. Despite certain evidence to the contrary in the dog (GLOYNA et al., 1970), the 5α-reductase activity in rat prostate is severely impaired during senescence or after castration (SHIMAZAKI et al., 1969), yet may be restored by the administration of testosterone, provided that RNA and protein synthesis can occur (SHIMAZAKI et al., 1970). The specificity of this induction of enzyme activity is considerable; a positive correlation was established between 5α-reductase activity and citric acid accumulation after androgenic stimulation whereas ATPase and 17β-hydroxysteroid dehydrogenase activities remained unchanged (SHIMAZAKI et al., 1969, 1970). Of these parameters, only citric acid is a classical marker for androgenic activity (MANN, 1964). In contrast to the prostate, the 5α-reductase activities in rat liver (BULLOCK et al., 1971) and adrenal (KITAY, 1968) are not influenced by changes in androgenic status, emphasizing again the unique properties of the enzyme in male accessory sexual glands. Polypeptide hormones do not influence 5α-reductase activity (MOORE and WILSON, 1973).

The testicular feminisation syndromes provide novel systems for investigating the relationship between androgen metabolism and the circulating concentrations of androgens. In the Stanley-Gumbeck pseudohermaphrodite rat, 5α-reductase activity persists to a significant extent in the remaining androgen target cells (BARDIN et al., 1970; BULLOCK et al., 1971; GROSSMAN et al., 1971) despite the absence of external male genitalia. In the clinical syndrome, as in experimental Tfm mutants (BARDIN et al., 1969), plasma levels of testosterone remain within the normal range (SOUTHREN et al., 1965; FRENCH et al., 1966). Despite a normal androgenic milieu, the general consensus of opinion is that measurement of 5α-reductase activity may be of outstanding value in the early diagnosis of testicular feminisation. This contention is supported by the widespread reports that 5α-reductase activity in skin from the mons pubis or labia majora of testicular feminisation patients is significantly lower than in similar skin samples from normal women or the pubic skin of men (GLOYNA and WILSON, 1969; HEINRICHS et al., 1969; NORTHCUTT et al., 1969). The only discordant report on this topic is that by PÉREZ-PALACIOS et al. (1971); these workers found similar levels of 5α-reductase activity in human skin, irrespective of the sex of origin or genotypic aberrations. Nevertheless, the general opinion is that a defect in 5α-reductase activity may provide a plausible explanation for certain forms of the testicular feminisation syndrome (WALSH et al., 1974). It is now possible to obtain viable cultures of fibroblasts from perineal skin and to screen them for 5α-reductase activity. Whereas 5α-reductase was readily detected in skin fibroblasts derived from normal men, it was totally undetectable in cultures prepared from the sexual skin of patients with familial incomplete male pseudohermaphrodite type 2 syndrome (WILSON, 1975; MOORE et al., 1975).

While the 5α-reductase system indubitably provides the majority of the intracellular 5α-dihydrotestosterone in secondary sexual glands, the supply of this active metabolite from peripheral sources cannot be ignored. Indeed, male plasma of experimental animals (ITO and HORTON, 1970; TREMBLAY et al., 1970) and man (ITO and HORTON, 1971) contains appreciable quantities of 5α-dihydrotestosterone. From a very detailed and dynamic analysis on the dog prostate TREMBLAY et al.

(1972) concluded that only one third of the 5α-dihydrotestosterone was formed within the gland from plasma precursors and the remainder was contributed directly by the testicular secretion. In certain species, therefore, the contribution of 5α-dihydrotestosterone from sources outside the androgen target cells can be very significant indeed.

Baulieu and his collaborators have forwarded the provocative concept that distinct biological responses are mediated by the various active metabolites of testosterone (BAULIEU et al., 1968). It was advocated that 5α-dihydrotestosterone specifically controlled cell division whereas 5α-androstan-3β,17β-diol regulated the secretory processes. The metabolism of testosterone in rat prostate maintained in organ cultures has been extensively investigated to validate this concept (ROY et al., 1972 a, 1972 b). These important studies indicated that the 5α-reductase system was essentially irreversible and in addition, that numerous active metabolites are necessary for the full expression of the androgenic response; this cannot be elicited by 5α-dihydrotestosterone alone. The full implications of these studies by ROY et al. (1972a, b) will be clearer when the specific effects of particular metabolites can be equated with biochemical rather than morphologic markers, but the detailed study conducted independently by GITTINGER and LASNITZKI (1972) strongly corroborates the concept that active metabolites have discrete biological functions. A corollary of this work is that classical bioassays for androgenic activity may need to be replaced by more sophisticated procedures. Certainly the "multimetabolite" concept is more attractive than other possible alternatives, especially that active metabolites stimulate a common biological mechanism but with varying degrees of effectiveness. The failure of cyproterone acetate to inhibit all the processes mediated by testosterone and its active metabolites militates against this latter postulate (MANGAN et al., 1973).

Metabolic inhibitors for the particular enzyme reactions involved in overall metabolism of testosterone would also be of great value, but few are currently available. The majority of antiandrogens impair the metabolism of androgens to only a limited extent; this is certainly the case with BOMT and flutamide (MANGAN and MAINWARING, 1972; MAINWARING et al., 1974 c). However, potential inhibitors of 5α-reductase include 17β-carboxyandrost-4-en-3-one (VOIGT and HSIA, 1973), 17β-hydroxy-16,16-dimethylestr-4-en-3-one (HUNT and NICHOLSON, 1974) and 6,17-dimethyl-4,6-pregnadien-3,20-dione (GIVNER and JAGARINEC, 1974). Selective inhibitors of hydroxysteroid dehydrogenases are also in the course of development (CHAPDELAINE et al., 1974).

3. Developmental Aspects of the Metabolism of Testosterone

Not only does the ability of androgen target cells to metabolise testosterone vary enormously, but major changes occur during foetal development (WILSON and LASNITZKI, 1971) and ageing (GLOYNA and WILSON, 1969; SHIMAZAKI et al., 1969; WILSON and WALKER, 1969). Modulation of metabolism is of great significance to the mechanism of action of androgens, especially from the developmental standpoint and this topic will be briefly reviewed.

a) Foetal Development

Classical embryologic studies, reviewed by WILLIER (1939), established that selected areas of the male reproductive tract originated from three embryonic anlages or primordial structures, as follows; the urogenital sinus (prostate and urethra), the urogenital tubercle (external genitalia), and the Wolffian duct (epididymis, vas deferens, and seminal vesicle). Androgens are necessary for the differentiation of these embryonic anlages but not for the regression of the Müllerian duct (for review, see JOST, 1970). Analyses on the foetal testis of sheep (BLOCK, 1964) and other species (LIPSETT and TULLNER, 1965; NOUMURA et al., 1966) confirmed that testosterone was secreted at a time commensurate with sexual differentiation. Elegant studies by JOSSO (1971, 1972, 1974) have indicated that the foetal testis also produces a polypeptide hormone that specifically initiates the dissolution of the Müllerian duct, but an understanding of the more complex involvement of androgens in the development of the male urogenital tract has been achieved only recently. In an outstanding contribution, demanding the highest technologic expertise, CATT et al. (1975) established that foetal gonadotrophins provided the hormonal stimulus for the maturation of the foetal Leydig cells and the onset of testosterone secretion. The increase in the binding of gonadotrophins in the foetal rabbit testis occurred precisely at the time when differentiation of the embryonic anlages was beginning (CATT et al., 1975). In several species, 5α-reductase activity was readily detected in the urogenital sinus and urogenital tubercle, suggesting that 5α-dihydrotestosterone was implicated in the differentiation of the prostate and external genitalia (WILSON and LASNITZKI, 1971; WILSON and SIITERI, 1973). In sharp contrast, however, 5α-reductase activity was not detected in the Wolffian duct at any developmental stage (WILSON and LASNITZKI, 1971; WILSON, 1973) suggesting that testosterone was the embryonic regulator for the differentiation of the epididymis, vas deferens and seminal vesicle. To corroborate this viewpoint WILSON (1973) also demonstrated a binding mechanism specific for testosterone in the developing Wolffian duct. Taken overall, these studies indicate the subtlety of androgen metabolism during foetal life and the implication of these developmental changes in the differentiation and maturation of the male phenotype. How metabolism is modulated in the various anlages is not known and androgen-binding mechanisms in the urogenital sinus and urogenital tubercle have not yet been described. In a particularly fascinating investigation CUNHA (1975) dissected out the urogenital sinus and vagina from female mouse embryos and transplanted them into neonatal male hosts of between 1 and 30 days of age. Both the embryonic sinus and vagina responded to the new androgenic milieu and developed into prostate glands in 1-day-old mice but not in recipients of 5 days of age and older. From exacting recombination experiments, CUNHA (1975) also demonstrated that the loss of susceptibility to androgens was attributable to changes specifically in the vaginal stroma rather than epithelium.

b) Neonatal Development

Although the centre of widespread interest and investigation, perhaps the most complex, yet least understood, of the developmental changes in androgen metabolism occur in the neonatal period. This contemporary interest stems from the widely held view that the secretion of androgens during a critical period immediately after birth controls the differentiation of both the hypothalamus-pituitary-gonad axis and certain target organs for steroid hormones (for review, see BARRACLOUGH, 1967). These changes evoked by androgens in the neonatal period are profound and irreversible. The study of these neonatal processes is an important aspect of neuroendocrinology, with extremely relevant applications to human syndromes with an endocrinologic aetiology. To illustrate recent findings in this absorbing area of current research, three particular aspects are briefly discussed.

It is well known that the growth of the seminal vesicles (MORRISON and JOHNSON, 1966; BRONSON et al., 1972) and penis (DIXIT and NIEMI, 1973) promoted by androgens during the attainment of sexual maturity in the adult can be markedly enhanced by prior administration of androgens during the neonatal period. As a continuation of this work, CHUNG and FERLAND-RAYMOND (1975) recently reported that rat seminal vesicle and coagulating gland, but not the prostate, were acutely sensitive to neonatal stimulation; in the terminology of these authors, the seminal vesicle may be androgenically programmed but the prostate may not. From both measurements of growth and the incorporation of [^{3}H] thymidine into DNA, CHUNG and FERLAND-RAYMOND (1975) made an important distinction between the response of the seminal vesicles to androgens during the neonatal or adult stages of development. Whereas the adult response to testosterone was inhibited by the antiandrogen, cyproterone acetate, that in the neonatal period was not. This suggests that the neonatal events are promoted by mechanisms other than the androgen receptor system or, more specifically, independent of the high-affinity binding of 5α-dihydrotestosterone. A lucid explanation of these neonatal phenomena is not yet available, but stroma-epithelium interactions may be involved (CUNHA, 1975). If this is indeed the case, then the stromal changes must be tissue-specific, thus occurring in the seminal vesicle but not in the prostate. It is also noteworthy in this regard that seminal vesicle and prostate originate from different embryonic anlages.

During the neonatal period, certain enzymes in the liver of the male rat are irreversibly destined to a strictly masculine trait by the secretion of testicular androgens (DE MOOR and DENEF, 1968; KRAULIS and CLAYTON, 1968; EINARSSON et al., 1973). This androgen-mediated phenomenon is generally referred to as neonatal imprinting, to distinguish certain hepatic enzymes with higher activities in male as compared with female rats. Imprinted enzymes originally included only steroid-reductases and dehydrogenases, but 7- and 16α-hydroxylases also fall into this category of sex-associated enzymes (TABEI and HEINRICHS, 1975). Regulation of these male-specific enzymes has recently been attributed to hormones from the hypothalamus, acting *via* the testes (DENEF, 1974; GUSTAFSSON and STENBERG, 1974a) and the adrenal (GUSTAFSSON and STENBERG, 1974b). In contrast to the findings on the neonatal effects of androgens in terms of seminal vesicle growth (CHUNG and FERLAND-RAYMOND, 1975), the imprinting of liver enzymes is sensitive to the anti-androgen, cyproterone acetate (GUSTAFSSON et al., 1975). This observation provides an important clue to the basic regulation of imprinting, namely an androgen receptor mechanism presumably specific for 5α-dihydrotestosterone.

A final aspect of the administration of androgens to neonatal rodents, but not rabbit (CAMPBELL, 1965) or Rhesus monkey, *Macaca mulatta* (TRELOAR et al., 1972), is the interference with the differentiation of hypothalamic centres which regulate somatic and psychic sexual function. In the newborn female rat, the administration of androgens provokes the anomalous development of the physical attributes and behavioural characteristics of the male. This process is termed neonatal androgenisation (for a review, see SAUNDERS, 1968). The administration of high doses of oestrogens at this critical developmental stage also produces dysfunction of normal female behaviour and anovulatory sterility results (GORSKI, 1963). It is generally considered that the small amounts of circulating oestrogens are prevented from reaching the neonatal brain by the presence of an oestrogen-binding protein in plasma, α-fetoprotein (VANNIER and RAYNAUD, 1975; DOUGHTY et al., 1975). This plasma protein is present in foetal and neonatal rats, but disappears some 29 days after birth (NUNEZ et al., 1971; RAYNAUD et al., 1971). On its disappearance, the increased secretion of oestrogens in the adult permits the normal female character to develop, largely due to a surge in the secretion of luteinising hormone (LH). Oestrogens exert a positive feedback control on the hypothalamus, resulting in a cyclical secretion of gonadotrophins; in the male and androgenised females, these pituitary hormones are secreted in a noncyclical manner. In the neonatal male, androgens are necessary for promoting the differentiation of the hypothalamus into the male pattern. At this critical neonatal stage of development, a single injection of oestrogens or androgens provokes permanent changes in enzymes outside the brain and liver, particularly in the kidney, adrenals, and gonads of both sexes (GHRAF et al., 1975).

At first sight, it is difficult to envisage how the neonatal administration of high doses of oestrogens or androgens can elicit similar phenomena, namely hypothalamic aberrations such as neonatal androgenisation and anovulatory sterility. Some insights into this complex aspect of brain function are now forthcoming and the findings are also probably germane to the process of enzyme imprinting. However, it should be emphasized from the outset that the explanation of neonatal androgenisation is by no means complete and considerable debate persists.

The first real clues came from a comparison of the effectiveness of various steroids on the differentiation of male sexual behaviour. LUTTGE and WHALEN (1970), BROWN-GRANT et al. (1971), and PAUP et al. (1972) found testosterone to be active in promoting neonatal androgenisation, whereas its esters or 5α-dihydrotestosterone were not. It has long been recognised that 5α-dihydrotestosterone, unlike testosterone, cannot be aromatised to oestrogens (MCGUIRE et al., 1960; ITO and HORTON, 1971) and the identification of an aromatising enzyme system in the anterior hypothalamus and limbic system offered an exciting explanation of neonatal androgenisation (KNAPSTEIN et al., 1968; NAFTOLIN et al., 1971; REDDY et al., 1974). The concept arose that conversion of testosterone to oestradiol-17β was the essential feature of the responses promoted by androgens in the neonatal brain and at first, seemingly convincing evidence supported the conversion concept. First, the aromatising enzyme system was conspicuously active in the neonatal brain rather than at other developmental stages (WEISZ and GIBBS, 1974). Second, oestradiol receptor systems with a high binding affinity were identified in the hypothalamus and limbic structures (PLAPINGER and MCEWEN, 1973; BARLEY et al., 1974; LIEBERBURG and MCEWEN, 1975; DAVIES et al., 1975). The specificity of these oestrogen receptors was unusual in permitting some binding of androgens; accordingly, the binding of oestradiol could be modulated by androgens (KORACH and MULDOON, 1975) and of particular importance, the neonatal effects of testosterone could be countered by an anti-oestrogen, MER-25 (MCDONALD and DOUGHTY, 1973). Third, while both testosterone and 5α-dihydrotestosterone could lower gonadotrophin secretion in weaned rats (SWEDLOFF et al., 1972; NAFTOLIN and FEDER, 1973), only testosterone was able to regulate pituitary secretions in the neonatal period (BROWN-GRANT et al., 1971; WHALEN and LUTTGE, 1971; MORISHITA et al., 1975).

On these collective observations, a persuasive explanation of neonatal androgenisation could be founded. The hypothalamus of the neonatal rodent could aromatise testosterone to oestradiol-17β within the brain itself, thus adroitly sidestepping the protective mechanism offered by α-fetoprotein. Thus, either the oestradiol formed from testosterone, or alternatively, massive doses of oestrogens administered directly could be specifically bound to the hypothalamus receptor system and thereby inhibit the release of gonadotrophins and other biological processes associated with the differentiation of the hypothalamus. If this conversion concept is essentially correct, the most emphatic support should be that 5α-dihydrotestosterone is unable to simulate the effects of testosterone, simply because it may not be aromatised. Certain limitations to this concept are now apparent. First, certain investigations indicate that androgens directly control the neural release of gonadotrophins rather than indirectly modifying their rate of secretion (KORENBROT et al., 1975). Second, recent work by GOTTLIEB et al. (1974) and GERRALL et al. (1974) seriously questions the contention that 5α-dihydrotestosterone is unable to evoke neonatal androgenisation. In the view of GOTTLIEB et al. (1974), the previous failures to counter the expression of female sexual characteristics with 5α-dihydrotestosterone may be simply attributed to its rapid degradation when administered conventionally in liquid injection vehicles. If, however, its biological life is extended by gradual release from subcutaneous pellets containing the plastic, Silastic 382 (Dow Corning Corporation), then indeed 5α-dihydrotestosterone can suppress the sexual differentiation of neonatal female hamsters (GOTTLIEB et al., 1974; GERRALL et al., 1975).

In summary, neonatal imprinting and androgenisation are complex phenomena, representing the most baffling examples of developmental changes in the metabolism of androgens. Despite the elegance of the conversion concept, certain aspects of the responses of the neonatal androgenisation of the hypothalamus still require unequivocal verification.

c) Adult Development

Acute changes in the metabolism of androgens also occur after birth. As a generalisation, all the male accessory glands of adult animals contain appreciable amounts of the enzyme, 5α-reductase (WILSON and GLOYNA, 1970) but exceptions to this premise are known. In dog prostate, 5α-reductase activity increases gradually during senescence (GLOYNA and WILSON, 1969), but adult rabbit and bull prostates have no 5α-reductase activity, yet the enzyme is present in both glands at birth and declines thereafter (GLOYNA and WILSON, 1969; WILSON and WALKER, 1971). A similar decrease in 5α-reductase associated with ageing has also been reported for skin (GOMEZ and HSIA, 1968; WILSON and WALKER, 1971). Perhaps the most complex developmental changes in 5α-reductase activity are evident in rat testis (INANO et al., 1967; FICHER and STEINBERGER, 1968). From a very high activity in the immature animal, testicular 5α-reductase activity declines progressively during development and is absent in the sexually mature adult. The secretion of 5α-dihydrotestosterone from the immature, but not the mature, rat testis has been confirmed by FOLMAN et al. (1972) by direct steroid analysis. Treatment of mature rats with hexoestrol restores testicular 5α-reductase activity (OSHIMA et al., 1967) whereas testosterone has the reverse effect (OSHIMA et al., 1970). A fine interplay between the 5α-reductase and testosterone clearly regulates the reductive metabolism of androgens in the testis. In prepubertal animals, testosterone synthesis is low yet testicular growth is maximal; at this developmental stage, 5α-reduced metabolites of testosterone are formed and these may be important in stimulating testicular growth. With sexual maturity, testosterone secretion is maximal and inhibits the testicular 5α-reductase. At this adult stage of development, testicular size is only maintained rather than stimulated and 5α-reduced steroids may not be required for cellular maintenance. Such a conclusion is harmonious with the views of other investigators in that 5α-dihydrotestosterone is principally involved in regulating mitosis (WILSON and GLOYNA, 1970). Recent work by DUBÉ et al. (1975) has demonstrated an even more subtle regulation of 5α-reductase activity in rat skin by steroid hormones. Both androgens and oestrogens modify enzyme activity but in characteristically distinct bodily sites. This interesting study probably indicates the full extent of the regulation of androgen metabolism by hormonal stimuli.

These results have a profound bearing on the developmental aspects of the mechanism of action of androgens, because the binding of testosterone and its active metabolites will not only be controlled by the presence or absence of a receptor mechanism, but also by the metabolic supply of biologically active steroids in both qualitative and quantitative terms. With reference to the developmental changes in 5α-reductase activity, the high-affinity binding of 5α-dihydrotestosterone has been reported in adult rabbit prostate and epididymis (MAINWARING and MANGAN,

1973; DANZO et al., 1973), adult rat testis (MAINWARING and MANGAN, 1973; HANSSON et al., 1974; MULDER et al., 1975) and human and rat skin (EPPENBERGER and HSIA, 1972; KEENAN et al., 1975). In all cases, binding of 5α-dihydrotestosterone *in vivo* will be limited not by the absence of receptor mechanisms, but by the limited provision of this active metabolite. The biological implications are threefold. First, receptors may be present in the adult animal but nonfunctional; second; receptors may be required only during acute developmental periods when organ growth is maximal and all active metabolites are freely available, and third, this surge of growth is probably coincident with the acquisition of sexual maturity.

d) Fluctuations in the Circulating Concentrations of Androgens

Thus far, only variations in androgen metabolism have been considered, but superimposed on these fluctuations are the changes in the secretion of testosterone by the testis. These variations are also important in the consideration of the developmental aspects of the mechanism of action of androgens.

In the human male, the secretion of testosterone rises abruptly at puberty and is maintained at a moderate level even in extreme old age (KENT and ACONE, 1966). Periodic fluctuations in the rate of testosterone secretion occur, however, even in normal individuals, with amplitudes of 14–42% in cyclical phases of between 8 and 30 days duration (DOERING et al., 1975).

Of the common experimental animals, the rat has been the most widely studied from the standpoint of variations in the androgenic milieu with age. Activation of a membrane-bound adenyl cyclase system by gonadotrophins clearly triggers the surge in testosterone secretion by the Leydig cells at the time of puberty (MENDELSON et al., 1975). Changes in the synthesis of 5α-dihydrotestosterone in the rat testis with development were described in detail in section II.3.c. Under conditions *in vitro*, testicular minces from adult animals metabolise endogenous or exogenous testosterone to 7α-hydroxylated derivatives, whereas similar preparations from immature animals form 5α-reduced metabolites of testosterone, especially 5α-androstanediols (LACROIX et al., 1975). After the attainment of sexual maturity, the plasma concentration of testosterone in the male rat gradually declines during senescence, while the much lower concentration of 5α-dihydrotestosterone, presumably derived from peripheral sources, is fully maintained (GHANADIAN et al., 1975). Recognition must also be given to the extremely local changes in the androgenic milieu as, for example, within the epididymal fluid; spermatozoa have an exceedingly high androgenic environment in the testis but during migration to the epididymis, this is reduced overall with a relative increase in the ratio of 5α-dihydrotestosterone to testosterone (VREEBURG, 1975). Cognisance must be given to all these factors when investigating the mechanism of action of androgens in relationship to development.

4. Androgenic Responses Not Mediated by 5α-dihydrotestosterone

As 5α-dihydrotestosterone has assumed such a critical place in the mechanism of action of androgens, and the review of steroid metabolism within androgen target cells has been primarily devoted to the enzyme, 5α-reductase, it is propitious to appraise critically the mandatory involvement of this active metabolite in androgenic responses. The involvement of 5α-dihydrotestosterone in androgenic responses is by no means invariant; from the summary presented in Table 3, it is evident that many biological responses may be evoked only by testosterone itself or metabolites other than 5α-dihydrotestosterone.

Certain aspects of Table 3 warrant further comment, especially in providing a foundation for later chapters on the biochemical processes involved in the manifestation of androgenic responses. Only testosterone is able to stimulate DNA synthesis in cultures of muscle cells (POWERS and FLORINI, 1975) and despite certain evidence to the contrary, androgen receptors in muscle appear to be specific for testosterone (JUNG and BAULIEU, 1972; MICHEL and BAULIEU, 1974). The biochemical effects of testosterone on cultured rat bone marrow cannot be mimicked by androgen metabolites, either with respect to stimulating RNA synthesis (SIERRALTA et al., 1974) or to retention within nuclear chromatin (MINGUELL and VALLADARES, 1974). Processes associated with haem synthesis are specifically enhanced by 5β- rather than 5α-metabolites of testosterone in chick blastoderm (LEVERE et al., 1967) and in cultures of foetal chick liver cells (GRANICK and KAPPAS, 1967). The specific effects mediated by testosterone in the brain were fully discussed in sections II.3.b. and II.3.c. The induction of β-glucuronidase in mouse kidney is exceedingly complex (see Chapter I.4.d and I.4.g), but it may be rapidly evoked by 5α-androstane-3β, 17β-diol rather than 5α-dihydrotestosterone (SWANK et al., 1973). Only testosterone can enhance glandular secretion in rat uterus (GONZALEZ-DIDDI et al., 1972) and this observation is consistent with the presence of a testosterone-specific receptor system (GIANNOPOULOS, 1973). In cultures of rat prostate, the regulation of the secretory process (BAULIEU et al., 1968) and many aspects of fine structure (GITTINGER and LASNITZKI, 1972) are selectively enhanced by metabolites other than 5α-dihydrotestosterone. Exclusively in canine prostate, the active metabolite is 5α-androstane-3α,17α-diol rather than 5α-dihydrotestosterone (EVANS and PIERREPOINT, 1975). The differentiation of the magnum in the chick oviduct requires a complicated synergism between testosterone and oestradiol (YU and MARQUARDT, 1973); this possibly is a testosterone-specific process, because 5α-reductase activity is low and certainly markedly below that in the adjacent oviduct structure, the shell gland (MORGAN and WILSON, 1970). From the work of TSCHOPP (1936) and HUGGINS et al., (1954), the growth and keratinisation of the immature rat vagina can be selectively promoted by Δ^5-androstenediol (3β, 17β-dihydroxyandrost-4-ene). In keeping with these studies, a receptor mechanism for Δ^5-androstenediol rather than testosterone or 5α-dihydrotestosterone is present in rat vagina (SHAO et al., 1975).

Table 3. Androgenic responses not mediated by 5α-dihydrotestosterone

Target cell	Biological response	References
Skeletal muscle	Anabolic response; growth	MICHEL and BAULIEU (1974) POWERS and FLORINI (1975)
Bone marrow	Stimulation of RNA synthesis	MINGUELL and VALLADARES (1974) SIERRALTA et. al. (1974)
Chick blastoderm Foetal chick hepatocytes	Stimulation of haem synthesis; induction of δ-aminolaevulinate synthetase	LEVENE et al. (1974) GRANICK and KAPPAS (1974)
Brain	Sexual behaviour and sex-specific characteristics	McDONALD et al. (1970) WHALEN and LUTTGE (1971 a)
Brain (neonatal females)	Anovulatory sterility	WHALEN and LUTTGE (1971 b)
Wolffian duct	Differentiation of the epididymis, seminal vesicle, and vas deferens	WILSON and LASNITZKI (1971) WILSON (1973)
Mouse kidney	Induction of β-glucuronidase	SWANCK et al. (1973)
Uterus	Stimulation of glandular secretions	GONZALEZ-DIDDI et al. (1972) GIANNOPOULOS (1973)
Rat prostate	Stimulation of glandular secretions	BAULIEU et al. (1968) ROY et al. (1972a, b) GITTINGER and LASNITZKI (1972)
Dog prostate	Enhanced DNA and RNA synthesis	EVANS and PIERREPOINT (1975)
Chick oviduct	Differentiation of the magnum	YU and MARQUARDT (1973)
Vagina (immature animals)	Growth and keratinisation	TSCHOPP (1936) HUGGINS et al. (1954) SHAO et al. (1975)

5. Interconversion of Steroid Hormones

Based primarily on the innovative studies by Samuels, Tait and their collaborators, it is now widely acknowledged that certain steroid hormones are freely interconvertible. The critical observation in the development of the "prehormone" concept was that conversion rates calculated from steroid analyses in blood differed from those estimated from urinary metabolites (BAIRD et al., 1969). Provided that a biologically inactive precursor of an active hormone may be found, then prehormone-hormone relationships may be extensively explored.

In the context of the mechanism of action of androgens, two metabolic interconversions predominate: androstenedione to testosterone (HORTON and TAIT, 1966) and androgen to oestrogen (LONGCOPE et al., 1969). Androstenedione is the principal androgen-related steroid secreted by the adrenal and hence may be an important source of active androgen metabolites, including 5α-dihydrotestosterone, other than their fabrication within androgen target cells. In certain species, the supply of 5α-dihydrotestosterone from peripheral metabolism may be very significant indeed (TREMBLAY et al., 1972). The practical implication of this observation is that experiments on certain species must be designed with care, because surgical castration need not necessarily deplete all potential sources of active androgen precursors. To emphasize this point, the administration of diethylstilboestrol is more effective than castration in suppressing the circulating concentrations of androgens in man (ROBINSON and THOMAS, 1971), presumably because it totally negates the secretion of pituitary gonadotrophins.

Metabolic transformation of androgens into oestrogens is also extremely important in the mechanism of action of androgens. The significance of hypothalamic aromatisation in the expression of neonatal phenomena was discussed earlier in section II.3.b. However, the implications of androgen-oestrogen interconversions may be of a far more general nature because the aromatising enzyme system is widely distributed, producing oestrone and oestradiol-17β from androstenedione and testosterone, respectively. In man, the aromatisation complex has been clearly identified in adipose tissue (SCHINDLER et al., 1972; NIMROD and RYAN, 1975), limbic structures and hypothalamus (NAFTOLIN et al., 1971), carcinomata (WOTIZ et al., 1955; KIRSCHNER et al., 1974), placenta (HORN and FINKELSTEIN, 1971), and, surprisingly, human hair (SCHWEIKERT et al., 1975). The enzyme complex is similarly demonstrable in a diversity of rat organs, including hypothalamus (REDDY et al., 1971), bone (VITTEK et al., 1974) and the dimethylbenzanthracene-induced mammary tumor (MILLER et al., 1974). The aromatising enzyme complex need not necessarily produce active oestrogens, however, because epitestosterone may be converted to the inactive stereoisomer, oestradiol-17α (HIGUCHI and VILLEE, 1970). The biological implications of this work are important. First, potent oestrogens may be formed from androgen precursors or prehormones in a variety of cell types. The prime advantage of intracellular synthesis over ovarian secretion is that the oestrogen is released in a free or active state, being protected from biological inactivation by plasma proteins, including the sex steroid-binding β-globulin, SBG (MURPHY, 1968) and particularly the fetal and neonatal estrogen-binding protein, α-fetoprotein (NUNEZ et al., 1971; RAYNAUD et al., 1971). Second, the conversion of androgens into oestrogens adds a new dimension of complexity to the mechanism of action of steroid hormones, because the biological responsives within putative androgen target cells may in reality be mediated by intracellular oestrogens. Substantive evidence for the latter phenomena could include neonatal androgenisation and imprinting (section II.3b) and the growth of hair in certain bodily sites in man (LISSER et al., 1974) and the rat (EPPENBERGER and HSIA, 1972).

6. Conclusions

Without question, intracellular metabolism has a profound influence on the mechanism of action of androgens, adding an aspect of complexity and biological subtlety not shared by other classes of steroid hormones. In the final analysis, the metabolic potential and capability of an androgen target cell will be of equal importance to the presence of androgen receptor systems in the overall expression of biological responses. Furthermore, it should always be borne firmly in mind that the activity of many enzymes is subject to modulating influences of development, differentiation, and the hormonal milieu. This is of critical importance in the planning and subsequent interpretation of experiments on androgenic regulation. It would also be naïve to assume that all androgenic responses are mediated by the intracellular formation and binding of 5α-dihydrotestosterone; it is now abundantly clear that certain biological phenomena are specifically regulated by testosterone itself, other active 5α-reduced metabolites, 5β-reduced metabolites, and even oestrogens.

III. Initial Events in the Mechanism of Action of Androgens

In the model for the mechanism of action of androgens presented in Figure 1 (page 9), particular importance was attached to the initial events in the androgenic response. This critical phase involves the high-affinity binding of androgens and it would appear that these initial interactions set in motion the broad spectrum of metabolic processes responsible for the manifestation of the hormonal response. Apart from the important provisos discussed in the latter part of Chapter III.1.g it is also reasonable to suggest that much of the specificity of androgenic responses is imbued in the binding characteristics of the androgen receptor system. On present evidence, the binding of active metabolites to receptor proteins is among the earliest detectable events occurring in the overall process of hormonal stimulation; only the entry and metabolism of testosterone can proceed faster. There appears to be no equivalent among androgen-responsive tissues to the remarkably rapid biochemical changes evoked in rat uterus by oestrogens (SZEGO and DAVIS, 1967). Other parameters are stimulated relatively rapidly by androgens, but it is difficult to make a convincing case that any of these possess the necessary specificity to discredit the paramount importance of the receptor system. It is more likely, if not certain, that these metabolic responses are also triggered by the high-affinity binding of androgen metabolites. The distinct steps in the selective binding of testosterone and its metabolites are presented in the sequence in which they probably occur in the intact androgen target cell. So much of the work on the androgen receptor mechanism was originally established in rat ventral prostate that a survey of these studies is presented first. The implications of these findings to other androgen responsive cells will then be summarised and the concluding sections of this chapter are devoted to the rapid stimulation of important metabolic processes by testosterone and its metabolites.

1. The Uptake, Retention, and Release of Androgens: Studies on Rat Ventral Prostate Gland

a) Transport of Testosterone and Related Steroids in Plasma

This topic was covered so extensively by Dr. U. Westphal in an earlier monograph in this series (Steroid-Protein Interactions) that only passing comment is necessary. It is now abundantly clear that all classes of steroid hormones are transported in the form of complexes with binding proteins in plasma (WESTPHAL, 1971) and both androgens and oestrogens are distributed by their high-affinity association with the sex steroid-binding protein, SBG. In certain animals, notably among members of the Rodentiae, SBG is absent (MURPHY, 1968) and plasma testosterone and 5α-dihydrotestosterone are transported by the corticosteroid-binding α_2-globulin, CBG. The association of testosterone with these plasma proteins averts indiscriminate androgenisation yet on the other hand, protects the hormone from premature destruction by the catabolic processes in the liver and elsewhere. An androgen-binding protein distinct from SBG is a feature of certain elasmobranchs, such as of the thorny skate, *Raja radiata* (IDLER and FREEMAN, 1968), but it is absent from the plasma of higher animals. From studies by TREMBLAY et al. (1972) and others (see Chapter II.2), the plasma of certain species of male animals contains appreciable amounts of 5α-dihydrotestosterone derived from peripheral sources. However, this active metabolite is bound with a higher affinity to SBG than testosterone (HORTON et al., 1967; KATO and HORTON, 1968) and this partly denigrates the biological importance of 5α-dihydrotestosterone directly available from the plasma. Certain androgen target cells, including nodules of human benign prostatic hyperplasia, are heavily infiltrated with plasma proteins, including SBG and CBG (STEINS et al., 1974; ROSEN et al., 1975). Consequently, very precise methods of analysis must be employed for distinguishing between the high-affinity binding of 5α-dihydrotestosterone to receptor proteins rather than CBG and SBG. Suitable methods include sucrose density gradient centrifugation (MAINWARING and MILROY, 1973; ROSEN et al., 1975) and agar gel electrophoresis (WAGNER and JUNGBLUT, 1976).

In the time since the reference work by WESTPHAL (1971) was published, certain important developments on the structure, function and analysis of the steroid-binding proteins in plasma have been reported. The purification of CBG (TRAPP et al., 1971) and SBG (MICKELSON and PÉTRA, 1975; ROSNER and SMITH, 1975) has been expedited by exploiting the extreme specificity of affinity chromatography. These proteins are now available in a purity approaching homogeneity and they will certainly be the first proteins with a high affinity for steroid hormones to be subjected to rigorous chemical analysis and even determinations of their amino acid sequence. They also provide suitable material for sophisticated physicochemical studies on the interactions between steroids and pure proteins (VAN BAELEN et al., 1972; STROUPE and WESTPHAL, 1975).

The biological importance of SBG has been substantiated in recent studies by LASNITZKI and FRANKLIN (1975). These authors have rightly emphasized that the concentration of SBG in plasma regulates the supply of testosterone and its active metabolites to the androgen target cells. Despite its high affinity for 5α-dihydrotestosterone, SBG cannot simulate one vital function of androgen receptor proteins, namely the transfer of this active metabolite into chromatin (IRVING and MAINWARING, 1973; MAINWARING et al., 1976). For this reason, SBG is the ideal control for studies on the rôle of androgen-receptor complexes in metabolic regulation.

Another extracellular binding protein for androgens is also worthy of particular mention, the androgen-binding protein, ABP, in epididymal fluid (FRENCH and RITZÉN, 1973; HANSSON et al., 1973; RITZÉN et al., 1973). From investigations on rats with experimental cryptorchidism, ABP is found to be a secretory product of the Sertoli cells in the testis (HAGENAS and RITZÉN, 1976) and passes by the efferent ducts into the lumen of the epididymis. ABP has been the focus of considerable research effort, but while it is important for the reproductive functions of the epididymis, it cannot be considered a receptor protein in this accessory sexual gland (TINDALL et al., 1975).

b) Entry of Testosterone into Cells

Surprising though it may seem, one of the least understood aspects of steroid hormones is their mode of entry into target cells. The plasma proteins prevent the indiscriminate release of androgens from the peripheral circulation (LASNITZKI and FRANKLIN, 1975) yet the small amounts of nonbound or active testosterone (VERMEULEN et al., 1971) were believed to enter cells by simple diffusion down a positive concentration gradient (for review, see KING and MAINWARING, 1974). Such entry by diffusion was originally contested by WILLMER (1961) and more recently by HEAP et al. (1970) who measured the passage of steroids through artificial membranes. However, it is now apparent that although steroids may enter all cells to a certain extent, their entry into steroid target cells may be preferentially facilitated. Specific transport mechanisms for the entry of steroid hormones have been described for rat uterus (ATGER, 1974), pituitary adenocarcinoma (HARRISON et al., 1975) and the prostate glands of dog and man (GIORGI et al., 1973, 1974; GIORGI, 1976). Utilizing the superfusion procedure developed by GURPIDE and WELCH (1969), the elegant studies by GIORGI and her collaborators provide the most penetrating insight into the dynamic relationships between testosterone and 5α-dihydrotestosterone and androgen target cells. Although seemingly not present in abundance, the surface carrier mechanism can actively transport amounts of testosterone and 5α-dihydrotestosterone far in excess of their concentrations in male plasma (GIORGI et al., 1973). Active transport speeded the uptake of testosterone down a positive concentration gradient and was the sole factor responsible for the entry of 5α-dihydrotestosterone into canine prostate samples because a negative concentration gradient exists for this active metabolite (GIORGI et al., 1974). The presence of the carrier mechanism was confirmed under conditions where the effects of the surface adsorption of steroids and the fluxes of intracellular water were obviated (GIORGI, 1976). It has also been established that oestradiol-17β and the antiandrogen, cyproterone acetate, interfered with the active transport of testosterone (GIORGI et al., 1973). The structural elements of the carrier mechanism have not been identified in prostate but they are reputed to be proteins in rat uterus (MILGROM et al., 1972). Aside from their importance in fundamental terms, these studies suggest a novel approach to the chemotherapy of prostatic disorders by countering the entry of androgens rather than their intracellular binding. It is not clear whether this facilitated entry mechanism exists in all androgen target cells but this would seem a reasonably safe assumption with respect to the male accessory sexual glands at least.

c) Cytoplasmic Receptors for Androgens

There is little value in extensively recapitulating the studies which led to the identification of the cytoplasmic receptor proteins for androgens in rat prostate; for a review, see KING and MAINWARING (1974). In brief resumé, MANGAN et al. (1968) reported the association of radioactivity with prostate chromatin after the administration of [^{3}H] testosterone *in vivo*. Prompted by the two-step binding mechanism proposed for oestrogens in uterus (JENSEN et al., 1968), a search for a cytoplasmic androgen receptor began and a protein essentially specific for the high-affinity binding of 5α-dihydrotestosterone was independently reported by MAINWARING (1969 a), FANG et al. (1969), and UNHJEM et al. (1969). Since 5α-dihydrotestosterone was also retained specifically in prostate nuclei (ANDERSON and LIAO, 1968; BRUCHOVSKY and WILSON, 1968; MAINWARING, 1969 b), the critical importance of this active metabolite was recognised for the first time. The emphasis of Chapter II was also founded on this important observation; further reference to metabolism will not be made in this chapter, but it should be emphasized that the androgen receptor and the enzyme, 5α-reductase, are physically distinct moieties (MAINWARING, 1970).

The general consensus of opinion is that the high-affinity binding mechanism in rat prostate cytoplasm is essentially specific for 5α-dihydrotestosterone. Both direct methods, based on the relative binding of numerous [^{3}H] steroids and indirect methods, based on the displacement of the binding of [^{3}H] 5α-dihydrotestosterone, have been employed. Insignificant binding was reported for androstenedione (MAINWARING, 1969 a), cortisol (MAINWARING, 1969 a), oestradiol-17β (BAULIEU and JUNG, 1970; MAINWARING, 1969 a) and numerous stereoisomers of 5α-androstanediols (BAULIEU and JUNG, 1970). Binding of testosterone was approximately 25% that of 5α-dihydrotestosterone (BAULIEU and JUNG, 1970).

Based on these rather tentative studies of steroid specificity, a far more penetrating appraisal of the relationship between steroid structure and receptor binding has since been undertaken, notably by LIAO et al. (1973 a) and also by SKINNER et al. (1975). The salient features of the admirable study conducted by LIAO et al. (1973 a) are summarised in Table 4. Of the naturally occurring steroids, only 5α-dihydrotestosterone was bound to any appreciable extent; testosterone was a poor competitor for these high-affinity binding sites. A 17β-hydroxyl group was a mandatory requirement for

Table 4. The relative affinity of steroids for the cytoplasmic androgen receptor of rat prostate.

Receptor was prepared from the 100,000 *g* supernatant fraction of prostate homogenates and purified further by precipitation with $(NH_4)_2SO_4$ and gel-exclusion chromatography on Sephadex G-25. Samples were incubated with [^{3}H] 5α-dihydrotestosterone in presence of an excess of nonradioactive competing steroids. Competition was calculated with reference to nonradioactive 5α-dihydrotestosterone. Data are taken from work of Liao et al. (1973 a) J. biol. chem. **248**, 6154.

Parent steroid	Derivative used as competitor (classified by substituent groups)	Relative competition indices (5α-dihydrotestosterone = 100
(a) 5α-Dihydrotestosterone	None; 5α-Dihydrotestosterone	100
	7α-Methyl	40
	17α-Methyl	110
	7α, 17α-Dimethyl	60
	7β, 17α-Dimethyl	<10
(b) Testosterone	None; testosterone	<10
	7α-Methyl	20
	7β-Methyl	<10
	17α-Methyl	10
	7α, 17α-Dimethyl	20
	7β, 17α-Dimethyl	<10
(c) 19-Nortestosterone (17β-hydroxy-estr-4-en-3-one)	None; 19-nortestosterone	90
	7α-Methyl	260
	17α-Methyl	120
	7α, 17α-Dimethyl	350
(d) Estratriene	2-Oxa-17β-hydroxy-estra-4, 9-dien-3-one	40
	2-Oxa-17β-hydroxy-estra-4, 9, 11-trien-3-one	320
	2-Oxa-17α-methyl-17β-hydroxy-estra-4, 9, 11-trien-3-one	380
	17α-Methyl-17β-hydroxy-estra-4, 9, 11-trien-3-one	230
(e) 5α-Androstane	None; 5α-androstane	0
	3-Oxo	0
	3, 17-Dioxo	0
	17β-Hydroxy	0
	3α-17β-Dihydroxy	0

specific binding, as indeed was the presence of a 3-ketone group. However, the rather surprising androgen-like activity reported by Zanati and Wolff (1972) for certain nonsteroidal compounds places doubt on the absolute necessity for a complete, six-membered A ring. The angular methyl group at the C-19 position is not a prerequisite for binding, as is evident from the effective competition by 19-nortestosterone for the receptor sites. Substitution with methyl groups provided some interesting derivatives, the 7α- and 17α-methyl steroids being particularly active competitors. Indeed, 7α, 17α-dimethyl-19-nortestosterone competed more avidly than 5α-dihydrotestosterone for the binding sites on the receptor. This is consistent with the extreme activity of these methylated steroids in many biological test systems (Segaloff, 1963; Campbell et al., 1963). The high-affinitv binding of the nortestosterone compounds was confirmed by direct binding studies with [^{3}H]-7α, 17α-dimethyl-19-nortestosterone (LIAO et al., 1973 a). In addition, the high binding affinity of estratriene derivatives was consistent with their high biological activity (FEYEL-CABANES, 1963; BUCOURT et al., 1970). With additional information from molecular models, LIAO et al. (1973 a) concluded that overall flatness and thickness in a steroid molecule were more important structural determinants for tight binding to the cytoplasmic receptor than a complex electronic structure surrounding the Δ^4 double bond in the A ring.

The survey conducted by SKINNER et al. (1975) substantiates the view that steroids with a planar geometric shape generally have a high affinity for the soluble receptor in rat prostate. However, close comparison with the work of LIAO et al. (1973 a) reveals certain defects in experimental design. In contrast to the results cited in Table 4, SKINNER et al. (1975) found active displacement even with 5α-androstanedione; unlike LIAO et al. (1973 a), however, the latter studies were not conducted at saturating concentrations of [^{3}H] 5α-dihydrotestosterone.

Taken overall, the structural requirements necessary for pronounced binding to the high-affinity sites on the receptor protein reported here closely resemble the features found in steroids with high biological activities in many bioassay procedures (see Chapter II.1). This provides important evidence of the unique involvement of the receptor system in the mechanism of action of androgens.

In earlier investigations on the binding of androgens in rat prostate, classical methods of protein fractionation were used to separate the complex formed between the cytoplasmic receptor and [^{3}H]-labelled steroid ligands. The methods included sucrose gradient centrifugation (MAINWARING, 1969 a; UNHJEM et al., 1969; FANG et al., 1969), gel-exclusion chromatography (MAINWARING, 1969 a), precipitation with ammonium sulphate (FANG et al., 1969; MAINWARING and PETERKEN, 1971; VERHOEVEN, 1975) and dextran charcoal adsorption (MAINWARING et al., 1974 a). A disquieting aspect of binding studies on the soluble extracts of rat prostate is the inordinately large amount of low-affinity or nonspecific binding. By contrast, receptor binding is of high affinity and specific. These terms are used synonymously in the current literature and this practice is perfectly acceptable. The assay of the cytoplasmic receptor in rat prostate has been facilitated by new and enterprising methods. A modified dextran charcoal procedure has the advantages of speed, sensitivity, and simplicity (BOESEL and SHAIN, 1974); nonspecific binding is assessed by the influence of nonradioactive steroid competitors. Selective precipitation of [^{3}H]-labelled receptor complex by a mixture of polyethylene glycol 600 and γ-globulin has been advocated by GROVER and O'DELL (1975). However, recent work from this laboratory (R. A. IRVING: unpublished work) indicates that this procedure is not as accurate as originally thought. Excessive amounts of [^{3}H] steroid may be trapped within the precipitate containing the receptor complex and this may only be removed by further washing on Whatman GF/A glass fiber discs. With this provision, the method of GLOVER and O'DELL (1975) has much to offer for routine analysis. A novel procedure using commercially available antibodies raised against steroid-protein conjugates has been proposed for the selective removal of low affinity, nonspecific binding proteins (CASTEÑADA and LIAO, 1975). This interesting method exploits the fact that steroids associated with receptors are enveloped within hydrophobic areas of the protein and hence inaccessible to the antibodies (LIAO et al., 1973 b; BELL and MUNCK, 1973). By contrast, steroids bound to plasma proteins are fully exposed and susceptible to immunological precipitation. This procedure is extremely flexible and will be widely used in the future. Recent work (W. I. P. MAINWARING: unpublished observations) also indicates the sensitivity and precision of isoelectric focusing for receptor complex analysis in small tubes used for conventional discontinuous electrophoresis (BEHNKE et al., 1975); many samples may be processed concurrently and receptor complexes are resolved from plasma complexes within 2 h. Methodologic problems in the assay of androgen receptors in rat prostate cytoplasm were ably reviewed by HØISAETER (1973); this paper is strongly recommended for general background information. An interesting proposal has recently been made by BONNE and RAYNAUD (1975); they recommend [^{3}H] methyltrienolone [1] for androgen receptor assays rather than [^{3}H] 5α-dihydrotestosterone. This synthetic androgen (R 1881) has two advantages; first, it binds with a high affinity to the receptor protein but not the SBG and second, it resists metabolism by prostate extracts.

Earlier suggestions that maximal binding of androgens occurs approximately 24–36 h after bilateral orchidectomy have recently been substantiated (GROVER and O'DELL, 1975); at this time, endogenous androgens are virtually depleted. The actual amount of cytoplasmic receptor protein is considerably influenced by castration (SULLIVAN and STROTT, 1973). Following a high level of binding when endogenous androgens are depleted, proteolysis and other degradative processes severely reduce the amount of androgen receptor, but after reaching this nadir, the amounts rise again in long-term castrated animals. These complex fluctuations are not fully understood, but they are not provoked by changes in the circulating concentrations of either androgens or pituitary hormones (SULLIVAN and STROTT, 1973). It has also been demonstrated that low concentrations of cytoplasmic androgen receptor are a feature of very old rats (SHAIN and AXELROD, 1973).

Considerable effort has been channelled into the physicochemical characterisation of the cytoplasmic androgen receptor complex. Bearing in mind the severe limitation that the receptor is not yet available in a highly purified form, its physicochemical properties are listed in Table 5; details on the plasma protein, SBG, and the epididymal protein, ABP, are included for comparison. All of these proteins bind 5α-dihydrotestosterone with a high affinity but only binding to the receptor is sensitive to cyproterone acetate; furthermore, only the androgen receptor protein can promote the transfer of 5α-dihydrotestosterone into chromatin (MAINWARING and IRVING, 1973). Other considerations aside, these distinctive features illustrate the unique biological function of the receptor protein. Just two additional properties of the receptor protein should be stressed. First, only the receptor protein complex is an oligomeric structure and hence dissociable under conditions of high ionic strength (BAULIEU and JUNG, 1970); on the other hand, SBG is somewhat heterogeneous until highly purified, but then migrates as a single protein monomer during denaturing conditions of electrophoresis (ROSNER and SMITH, 1975). Second, only the receptor protein complex is modified by warming at 20° C to a smaller form with a different pI and $S_{20,\omega}$ (MAIN-

---

[1] Footnote: methyltrienolone is R1881 or 17β-hydroxy-17α-methyl-estra-4,9,11-trien-3-one.

Table 5. A comparison of the physical and chemical properties of androgen-binding proteins.

Property	(a) Cytoplasmic receptor, rat prostate		(b) Sex-steroid binding globulin, human plasma		(c) Androgen-binding protein, rat epididymis	
	Value	Reference	Value	Reference	Value	Reference
Molecular weight	276,000	Mainwaring (1969a)	52000	Mercier-Bodard et al. (1970)	90000	Hansson (1972)
			100000–110000	Gueriguan and Pearlman (1968)		
			98000	Corvol et al. (1971)		
Einstein–Stokes radius (Å)	84	Mainwaring (1969a)	47	Gueriguan and Pearlman (1968)	47–48	Hansson (1972)
			29.5	Corvol et al. (1971)		
Sedimentation coefficient (S)	8.0	Mainwaring (1969a)	4.6 [a]	Rosner and Smith (1975)	4.6	Hansson (1972)
S value in 0.5M KC1	4.5	Baulieu and Jung (1970)	Not determined		4.6	Hansson (1972)
Fractional ratio [b] (f/f$_0$)	1.96	Mainwaring (1969a)	1.87 [a]	Rosner and Smith (1975)	1.61	Hansson (1972)
Isoelectric point (pI)	5.8	Mainwaring and Irving (1973); Tindall et al. (1975)	6.6 [c]	van Baelen et al. (1969)	4.6–4.7	Hansson (1972); Tindall et al. (1975)
			5.5 [a]	Rosner and Smith (1975)		

Requirement for -SH groups	Yes	Mainwaring (1969a)	Yes	Pearlman et al. (1969)	No	Tindall et al. (1975)
Thermal stability	Extremely labile	Baulieu and Jung (1970)	Stable at 45° C	Vermeulen and Verdonck (1968)	Stable at 50° C	Tindall et al. (1975)
Dissociation constant [d] (K_d)	2.4×10^{-9} M 4.0×10^{-9} M	Ritzén et al. (1971) Mainwaring (1973)	1.1×10^{-9} M [e]	Mercier-Bodard and Baulieu (1968)	5.5×10^{-9} M	Hansson (1972)
Effect of cyproterone acetate [d]	Active competitor	Fang and Liao (1969)	No competition	Guillemant et al. (1969)	No competition	Tindall et al. (1975)
Relative binding of oestradiol [d]	Low	Baulieu and Jung (1970)	High	Mercier-Bodard and Bauleieu (1968)	Low	Hansson (1972)
Transfer of 5α-dihydrotestosterone into chromatin [f] and DNA [g]	Yes [f,g]	Mainwaring and Irving (1973)	No [f]	Mainwaring and Irving (1973)	No [h]	Tindall et al. (1975)
Effect of warming at 20° C	Modified form; [i] 4.5S, pI 6.5	Mainwaring and Irving (1973)	No change	Vermeulen and Verdonck (1968)	No change	Tindall et al. (1975)

[a] SBG almost homogeneous; [b] partial specific volume assumed as 0.725 cm^3/g; [c] microheterogeneous unless treated with neuraminidase; [d] for binding of 5α-dihydrotestosterone; [e] recalculated from original data for testosterone; [f] chromatin; [g] DNA; [h] by inference only; [i] higher binding to chromatin.

WARING and IRVING, 1973); the activated form binds more rapidly to the acceptor sites in nuclear chromatin. The implications of this transformation process are discussed at greater length in section III.1.f.

From perceptive studies performed by LIAO et al. (1971) and FANG and LIAO (1971) it is now clear that the receptor protein is not the only high-affinity binding component for androgens in rat prostate cytoplasm. An additional α-protein is present as well as the receptor or β-protein. These proteins differ fundamentally in their binding character and biological function. 5α-Dihydrotestosterone binds to the β-protein (receptor) to form complex II; it also binds to α-protein forming complex I. Only complex II is retained by prostate nuclei (FANG and LIAO, 1971) and is clearly the protein responsible for the nuclear retention of 5α-dihydrotestosterone. The α-protein antagonises this nuclear binding and no equivalent has been reported in other hormone-sensitive systems. The β-protein has an almost exclusive affinity for 5α-dihydrotestosterone and this binding is sensitive to cyproterone acetate; in marked contrast, the α-protein binds progesterone and oestradiol to an appreciable extent yet it is relatively insensitive to the antiandrogen. These interesting studies indicate that a considerable measure of regulation exists in the prostate cytoplasm for the overall transfer of 5α-dihydrotestosterone into the nucleus; clearly an important interplay exists between the α- and β-proteins. The only facet of this work which remains to be clarified is the relative ratio of these proteins in the intact prostate cell. Despite a similar sedimentation coefficient of 3.5S, the α- and β-proteins may be separated by $(NH_4)_2SO_4$ fractionation and gel-exclusion chromatography (LIAO et al., 1971).

Thus far, consideration has been given only to the binding of 5α-dihydrotestosterone in the soluble cytoplasm (or cytosol) of rat prostate. Work by BAULIEU et al. (1971) and ROBEL et al. (1974) has now provided irrefutable evidence that the microsomal fraction of rat prostate but not rat liver is capable of binding 5α-dihydrotestosterone and 5α-androstane-3β,17β-diol with a high affinity. There appears to be a single class of high-affinity binding sites for these two active metabolites, 5α-dihydrotestosterone (K_a 2.5 × 10^{10} M) and 5α-androstane-3β,17β-diol (K_a 6.5 × 10^{10} M); testosterone and the closely-related stereoisomer, 5α-androstane-3α,17β-diol, are weak competitors for these specific binding sites. Representatives of other classes of steroid hormones such as oestradiol, progesterone, and cortisol are not bound to any measurable extent. These binding sites on the membranes of the endoplasmic reticulum demonstrate steroid- and tissue-specificity, yet they are unequivocally distinct from the soluble androgen receptor protein. These extranuclear sites are relevant to the specific regulation of the secretory function of rat prostate as envisaged by BAULIEU et al. (1968) and GITTINGER and LASNITZKI (1972). In a wider context, microsomal binding is germane to the androgenic responses mediated in the cytoplasmic compartment by essentially translational mechanisms (see Chapter I.4.g). The dual specificity of microsomal binding, for example, possibly explains why both 5α-dihydrotestosterone and certain 5α-androstanediols can induce β-glucuronidase in other androgen-sensitive organs, such as mouse kidney (SWANCK et al., 1973). It is to be hoped that these important studies will be extended in the future, with particular attention being given to the solubilisation and characterisation of these microsomal binding proteins; one should perhaps hesitate in naming them as receptor proteins on current evidence.

d) Antiandrogens

The importance of androgen antagonists or antiandrogens in the development of the receptor concept was emphasized in Chapter I. Many antiandrogens, such as cyproterone acetate and BOMT, have a steroid-related structure and mainly elicit their biological antagonism by competing for the binding sites on the receptor protein (FANG and LIAO, 1969; MANGAN and MAINWARING, 1972). However, cyproterone acetate has a certain progestational activity and hence may express its antiandrogen function in more than one way (see, for example, GIORGI, 1976). Certain antiandrogens have no apparent semblance of steroid structure and the most active of these is flutamide or 4′-nitro-3′-trifluoromethylisobutrylanilide (NERI et al., 1972). Among its many features, flutamide is capable of suppressing prostatic hyperplasia in dogs (NERI and MONAHAN, 1972). Despite a nonsteroidal structure (see Fig. 3.), certain flexibility is present in the bonds of the side chain of flutamide and under conditions *in vivo,* there is no doubt that this antiandrogen counters the binding of 5α-dihydrotestosterone to the receptor system of rat prostate (MAINWARING et al., 1974; PEETS et al., 1974; LIAO et al., 1974). In all these publications and especially that by MAINWARING et al. (1974 a), there was negligible competition by flutamide

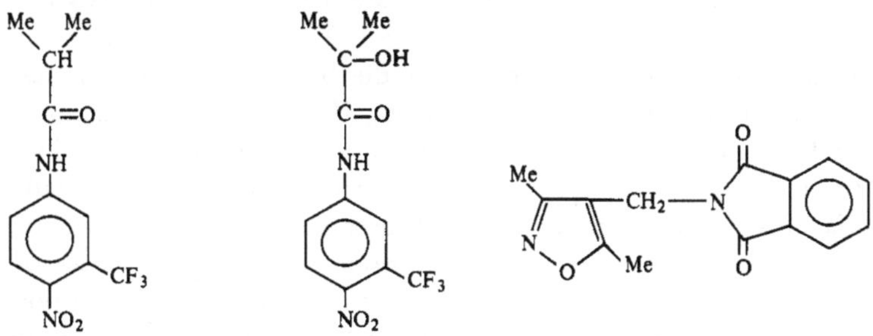

Flutamide[a,b] Flutamide metabolite[a] DIMP[a]

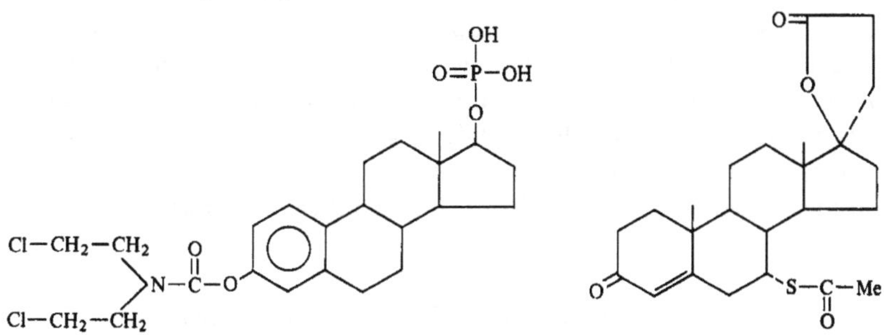

Estracyt Spirolactone SC 9420

Fig. 3. Structures of certain antiandrogens. [a] These compounds have negligible androgenic, estrogenic, or progestational activity; [b] *Me* methyl group

for the cytoplasmic receptor sites under conditions *in vitro*. On such evidence, it was predicted that flutamide needed to be converted to another compound for expression of its androgen antagonism. From the recent study by KATCHEN and BUXBAUM (1975), this certainly seems the case. Flutamide was rapidly converted in man to the related compound, α,α,α-trifluoro-2-methyl-4'-nitro-m-lactotoluidide by hydroxylation of the side chain and the high content of this metabolite in plasma suggests that it may be the active principle. Another nonsteroidal antiandrogen is DIMP or N-(3,5-dimethyl-4-iso-oxazoylmethyl) phthalimide (BORIS et al., 1973), but it has not been extensively studied at a molecular level and its mode of antagonism is not known.

The ideal antiandrogen should satisfy three essential criteria; low toxicity, negligible intrinsic hormonal activity, and a high affinity for the receptor mechanism. On current evidence, flutamide and its metabolite meet these demands and studies on their efficacy in the treatment of human prostatic carcinoma are awaited with interest. Implicit in the receptor hypothesis is a close geometric fit between the steroid ligand and the binding site on the receptor protein. As discussed earlier, binding of androgens demands a relatively planar structure and by flexibility in its side chain, flutamide can adopt the required configuration of a flat molecule (LIAO, 1976).

In the clinical context, it should not be overlooked that oestradiol-17β and diethylstilboestrol are potent antiandrogens (HUGGINS and HODGES, 1941). In the hope of raising its antiandrogenic activity, or more specifically, antitumour activity, oestradiol derivatives with a cytostatic nitrogen mustard side chain have been developed; the best known of these is Estracyt (see Fig. 3.). This is an interesting compound but despite the fact that Estracyt can impair the binding of 5α-dihydrotestosterone to cytoplasmic androgen receptors (HØISAETER, 1974), it was only marginally superior to diethylstilboestrol in arresting the growth of human prostatic carcinoma and also gave rise to cardiovascular episodes (MÜNTZING, et al., 1974). A detailed survey of the antiandrogenic effects of oestrogens has been presented by TVETER (1974). Inhibition of gonadotrophin secretion is the major antiandrogenic effect of oestrogens (HARNS and DONOVAN, 1966) but other aspects include the inhibition of the 5α-reductase system (SHIMAZAKI et al., 1965; LEAV et al., 1971) and a stimulation of the synthesis of SBG (POHLMAN et al., 1969). Oestradiol may also compete to a certain extent for the binding sites for 5α-dihydrotestosterone (see section III.1.e.). Together, these mechanisms counter the secretion and metabolism of testosterone. Spirolactone SC 9420 (17-hydroxy-7-mercapto-3-oxo-17α-pregn-4-en-21-carboxylic-γ-lactone-7-acetate) is an aldosterone antagonist widely used in the treatment of hypertension. As a deleterious side effect, the drug stimulates gynecomastia (STEELMAN et al., 1969) and it is now known that SC 9420 competes for the binding sites on the androgen receptor of rat prostate (BONNE and RAYNAUD, 1974; CORVOL et al., 1975). On this evidence, certain spirolactones are nonsteroidal antiandrogens.

e) The Nuclear Binding of 5α-dihydrotestosterone

From the historical standpoint, the selective binding of 5α-dihydrotestosterone in the nuclei of rat prostate, independently reported by ANDERSON and LIAO (1968), BRUCHOVSKY and WILSON (1968), and MAINWARING (1969b), was largely responsible for the resurgence of interest in the mechanism of action of androgens. The high-affinity binding of this active metabolite is a unique feature of many androgen-sensitive cells and remains the cornerstone of the current model for the mechanism of action of androgens (see Fig. 1.). An important aspect of the nuclear binding of [^{3}H] 5α-dihydrotestosterone in prostatic minces *in vitro*, first reported by ANDERSON and LIAO (1968) and later substantiated by MAINWARING (1970 b), was the temperature dependence of the process. Nuclear binding of 5α-dihydrotestosterone could not occur at 0° C, yet it proceeded at temperatures above 20° C and was maximal at 37° C.

Because of its extreme importance, the steroid specificity of the nuclear binding process in rat prostate has been extensively studied. Representative results are presented in Table 6 and it is clear from this summary that the specificity with respect to 5α-dihydrotestosterone is very marked indeed. No competition was evident with glucocorticoids, mineralocorticoids, or the biologically inactive stereoisomer, 5β-dihydrotestosterone (MANGAN and MAINWARING, 1972). Androstenedione and progesterone competed to a certain extent, but this may be attributed to their conversion to 5α-dihydrotestosterone under these experimental conditions (see also BRUCHOVSKY, 1972; BURIC et al., 1972). The oestrogens, stilboestrol, and oestradiol-17β, were relatively weak competitors and effected displacement only when present in very considerable excess (FANG et al., 1969). The steroid-related antiandrogens, cyproterone acetate and BOMT were also inhibitors of the binding of 5α-dihydrotestosterone (MANGAN and MAINWARING, 1972) but only at high concentrations. This explains why both antiandrogens must be administered in high doses in order to elicit their androgen antagonism *in vivo* (MAINWARING et al., 1973; RENNIE et al., 1975).

After the administration of [^{3}H] testosterone to castrated rats in vivo, autoradiography of tissue slices has been useful in corroborating the nuclear localisation of androgens. TVETER and ATTRAMADAL (1969) used freeze-drying and osmium tetroxide fixation in the preparation of their specimens and reported that silver grains were predominantly located in the nuclei of the epithelial cells; few grains were deposited in the underlying stroma or the glandular lumina. Utilising the more sophisticated "dry mount" technique which prevents any diffusion of water-soluble intracellular constituents SAR et al. (1970) produced autoradiographic pictures of the highest quality and confirmed the restricted association of [^{3}H]-labelled androgens within epithelial nuclei. This work emphasizes the importance of a multidisciplinary approach to biological problems, especially in providing confirmation of concepts based originally on biochemical procedures.

Before a detailed physicochemical characterisation of the nuclear receptor-5α-dihydrotestosterone complex could be entertained, it was necessary to find a means of releasing it from nuclei labelled *in vivo* or *in vitro* with [^{3}H]-testosterone or 5α-dihydrotestosterone. After unsuccessful attempts using freezing and thawing, surface-active detergents (Triton X-100 and Tween 25), and sodium deoxycholate, it has now become standard practice to release the nuclear receptor complex by extraction with 0.3–0.4 M KCl (BRUCHOVSKY and WILSON, 1968; FANG et al., 1969; MAINWARING, 1969 b). It should be stressed, however, that a significant proportion of nuclear-bound [^{3}H] 5α-dihydrotestosterone resists extraction even in 1.0 M KCl and the biological significance of this residual active metabolite remains in question (MAINWARING, 1969 b; MESTER and BAULIEU, 1972).

From studies on nuclear extracts in 0.4 M KCl, many properties of the nuclear

Table 6. The steroid specificity of the nuclear binding of 5α-dihydrotestosterone in rat prostate.

Binding of [^{3}H] testosterone and [^{3}H] 5α-dihydrotestosterone was measured in prostate minces at 37° C; after homogenisation, binding of [^{3}H] steroids was measured directly in purified nuclei or by sucrose gradient analysis of soluble nuclear extracts in 0.5M KCl. Replicate experiments were carried out in presence of excess of nonradioactive competitors. In experiments with [^{3}H] testosterone, between 78 and 87% of nuclear-bound radioactivity was recoverable as 5α-dihydrotestosterone.

Nonradioactive competitor	[^{3}H] 5α-Dihydrotestosterone [a]		[^{3}H] Testosterone [b]		[^{3}H] 5α-Dihydrotestosterone [b]	
	Ratio to [^{3}H] steroid	Binding	Ratio to [^{3}H] steroid	Binding	Ratio to [^{3}H] steroid	Binding
None		100		100		100
Testosterone			10	6		
Androstenedione	500	51	10	18		
5α-Dihydrotestosterone	200 500	12 11	10	25		
5β-Dihydrotestosterone	500	90				
Cortisol	500	97	10 2800	100 100		
Corticosterone	500	98				
Aldosterone	500	93				
Progesterone	500	42	10	58		
Diethylstilbestrol			10 2800	65 84	700 2800	71 81
Oestradiol-17β	500	72				
Cyproterone acetate [c]	40 200 400	44 41 27			400	30
BOMT [c]	40 200 400	51 44 28				

[a] from MANGAN and MAINWARING (1972); [b] from FANG et al. (1969); [c] steroid-related antiandrogens (see III.1.d).

receptor complex have been established. Determinations of sedimentation coefficient have proferred values of 3S (FANG et al., 1969; BAULIEU et al., 1971) and 4.5S (MAINWARING, 1970 b; MAINWARING and PETERKEN, 1971); this disparity is probably attributable to methodologic differences between laboratories. From gel-exclusion chromatography studies, the molecular weight is judged to be approximately 100,000 (BRUCHOVSKY and WILSON, 1968; MAINWARING, 1969 b; UNHJEM, 1970). The nuclear receptor complex has a marked tendency to precipitate as dense aggregates under conditions of low ionic strength but provided it is freed of basic protein contaminants, the complex gives a single peak, pI 6.5, in isoelectric focusing (MAINWARING and IRVING, 1973; IRVING and MAINWARING, 1974). The solubilised receptor complex is stable at 4° C for protracted periods but is rapidly degraded at temperatures of 45° C and above (FANG et al., 1969; MAINWARING, 1970 a); it also has a pronounced instability at pH values below 6 or above 9.5 (FANG et al., 1969). From this summary, it is abundantly clear that the nuclear receptor complex differs markedly in physicochemical properties from the soluble (cytoplasmic) androgen receptor complex (compare Table 5.). If the high molecular weight form (8S, pI 5.8) constitutes the actual form of the cytoplasmic receptor in the intact cell, then profound changes occur in its structure during translocation of the active metabolite of testosterone into nuclear chromatin.

f) Reconstituted, Cell-Free Systems for the Nuclear Transfer of 5α-dihydrotestosterone: the Acceptor Hypothesis

It is now pertinent to appraise the involvement of the cytoplasmic androgen receptor complex in the overall binding process, resulting in the nuclear localisation of 5α-dihydrotestosterone. At first, the mandatory involvement of the cytoplasmic receptor protein was difficult to visualise, nor indeed did it seem necessary for nuclear binding. First, with a partial localisation of 5α-reductase in nuclei (BRUCHOVSKY and WILSON, 1968), it could be argued that 5α-dihydrotestosterone was formed directly within the nucleus from testosterone reaching the nuclear membrane simply by diffusion. Second, NEAL (1970) claimed that 5α-dihydrotestosterone became firmly associated with purified nuclei in the absence of cytoplasmic proteins *in vitro* and under similar conditions, BASHIRELAHI and VILLEE (1970) reported that certain metabolic processes in nuclei could be stimulated directly by androgens *in vitro*. These reports ran counter to other observations that indicated the necessity of the cytoplasmic androgen receptor protein. MAINWARING (1969 b) found that many steroids, including androgens, could be bound directly to soluble extracts of prostate nuclei but the binding was nonspecific and of exceedingly low affinity. In similar studies, FANG et al. (1969) reported that although androgens could be retained on incubation of highly purified nuclei alone, they could not be extracted in a form yielding a single, discrete peak in sucrose density gradients; they concluded that such binding was nonspecific. The most graphic demonstration of extensive cytoplasmic binding prior to nuclear binding came from the study by MAINWARING and PETERKEN (1971); they determined the subcellular distribution of radioactivity in the prostate gland at various times after the administration of a single dose of [^{3}H] testosterone, *in vivo*. Binding of radioactivity in the microsomes but most conspicuously in the cell soluble fraction preceded nuclear binding. Insignificant binding was measured in the mitochondria at any time, but 30 min after the injection of [^{3}H] testosterone, the nuclei essentially accounted for all the radioactivity bound with a high affinity in the prostate gland (Fig. 4.). Over 80% of the tritium associated with the cytoplasmic and nuclear fractions was recovered as [^{3}H] 5α-dihydrotestosterone.

On these firmer foundations, creditable attempts were made to simulate the transfer of 5α-dihydrotestosterone into chromatin or other nuclear preparations in reconstituted, cell-free systems *in vitro*. The results of these investigations are summarised in Table 7. Four important facts emerge from these studies. First, with the notable exception of the work of NOZU and TAMAOKI (1975), the transfer of 5α-dihydrotestosterone into nuclei had an absolute requirement for the receptor protein; this function could not be mimicked by rat plasma (MAINWARING and PETERKEN, 1971) or human SBG (MAINWARING and IRVING, 1973). Second, in keeping with nuclear binding in intact prostate tissue (ANDERSON and LIAO, 1968; MAINWARING, 1970 b), transfer into nuclei was a temperature-dependent process which could not occur at temperatures below 20° C. Temperature dependence was evident with transfers into chromatin (MAINWARING and PETERKEN, 1971), but not in the association of 5α-dihydrotestosterone with soluble nuclear constituents (TYMOCZKO and LIAO, 1971; LIAO et al., 1972). Third, evidence of the saturation of nuclear sites was reported in studies on chromatin (MAINWARING and PETERKEN, 1971; STEGGLES et al., 1971) but not purified DNA. Collectively, this evidence suggested that although DNA may be implicated in the overall binding process, specificity lies in the nuclear-associated proteins. Specificity in the source of DNA was claimed in the anomalous report by CLEMENS and KLEINSMITH (1972) on oestrogen receptor complexes; this has never been seen in studies on androgen target cells or indeed by others working on rat uterus (KING and GORDON, 1972). Finally, it was reported by most research teams that nuclear preparations from androgen-target cells were more efficient in the retention of the cytoplasmic androgen receptor complex than similar preparations from nontarget cells. This prompted the formulation of the acceptor hypothesis by LIAO et al. (1972). They suggested that as a result of differentiation or hormonal stimulation, the chromatin of androgen target cells contains a higher complement of the acceptor components responsible for the retention of the cytoplasmic receptor complex. The precise nature of the acceptor sites is discussed in section III.1g.

At the end of section III. 1e, differences in the physicochemical properties of the cytoplasmic and nuclear forms of androgen receptor complexes were highlighted. In addition, it is now clear that the nuclear binding not only requires the presence of the cytoplasmic (soluble) receptor complex but is also temperature-dependent. The vexatious question, then, is how may these observations be reconciled? Evidence accumulated from many steroid-sensitive systems indicates that the cytoplasmic receptor complex undergoes a transformation or activation step prior to its association with nuclear components; this process was reviewed by KING and MAINWARING (1974) but the extensively studied oestrogen receptor system is included here merely to illustrate the basic aspects of receptor transformation. While there is still some doubt as to the structure of the cytoplasmic oestrogen receptor complex *in vivo*, and also whether it is composed of identical subunits or not, it is universally agreed that the 8S form of the receptor has an oligomeric structure. Consequently, the 8S receptor may be dissociated into several forms containing various combinations of subunits,

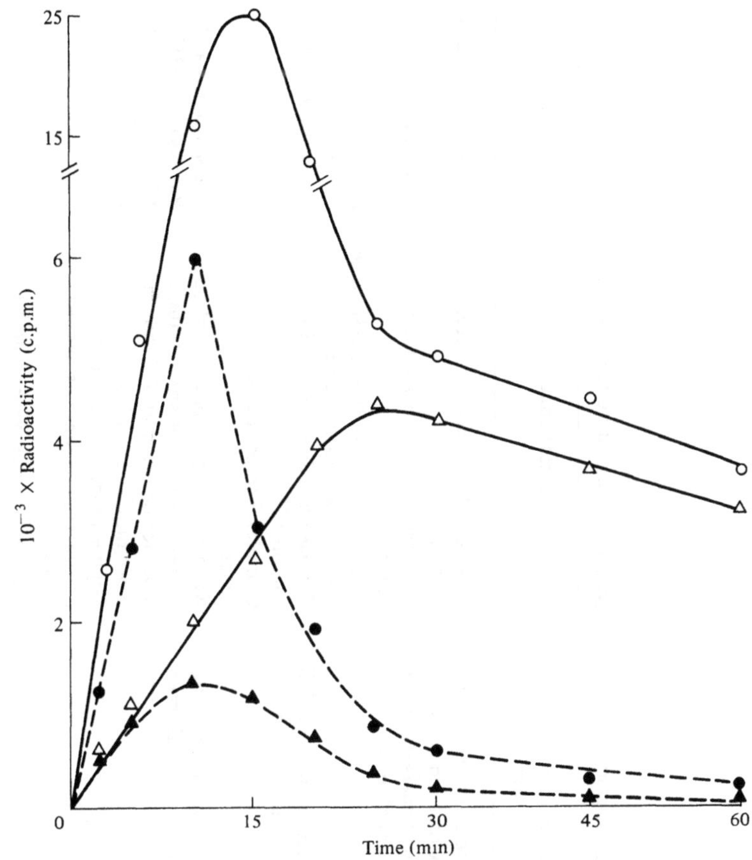

Fig. 4. The intracellular distribution of radioactivity in the rat prostate after a single subcutaneous injection of [^{3}H] testosterone *in vivo.* Beginning 24 h after castration, animals received injection of 10.6 μCi of ^{3}H testosterone and at various times thereafter, glands pooled from two animals were homogenised in sucrose-EDTA medium and separated into subcellular fractions by differential centrifugation, as described by HARDING and SAMUELS (1961). Bound radioactivity in cytoplasmic fraction was precipitated with protamine sulphate; only receptor-bound radioactivity is measured by this polycationic compound (MAINWARING, 1969 a). Data are based on work of MAINWARING and PETERKEN (1971). Biochem. J. 125, 285. Total radioactivity, ○; cytoplasm (bound only), ●; nuclei, △; microsomes, ▲

including a 4.5S species. The complex extracted from nuclei in 0.5 M KCl has a similar but nevertheless distinct sedimentation coefficient of 5S. While rigorous proof is not available, the general consensus of opinion is that a 4.5S portion of the 8S receptor constitutes the functional moiety of the 5S nuclear form of receptor. The formation of the 5S form of receptor was demonstrated in cytoplasm alone by BRECHER et al. (1971) and MOHLA et al. (1972); the transformation process required the presence of oestradiol-17β, a physiological pH, and elevated temperatures near 30° C. Observations by ROCHEFORT (1971) and particularly PUCA et al. (1972) suggest that a protease or other activating factors may be necessary for transformation of cytoplasmic receptor complexes.

Table 7. Reconstituted cell-free systems for studying the association of $[^3H]$ 5α-dihydrotestosterone with nuclear components.

Nuclear components present	Source of receptor	Assay of bound $[^3H]$ steroid	General comments	References
(1) Purified nuclei	(a) Unfractionated cytoplasm	Extraction of nuclei with 0.5M KCl and sucrose gradient analysis; washing on glass fibre discs [a]	Absolute dependence on receptor proteins. Transfer occurred at 20° C or 37° C, but not at 0° C. Higher binding in target cell nuclei (prostate as against liver).	ANDERSON and LIAO (1968) FANG and LIAO (1969) MAINWARING (1970 b)
	(b) Fractionated cytoplasm; $(NH_4)_2SO_4$ fractions α- and β [b]	Extraction in 0.5M KCl and sucrose gradients	Transfer mandatorily required complex II [b] Complex I impaired the transfer process.	FANG and LIAO (1969)
(2) Purified chromatin	(a) Fractionated cytoplasm; receptor-enriched fraction or 5000-fold purified receptor	Collection and washing of chromatin on Millipore membranes	Required receptor protein; plasma proteins could not affect transfer. Transfer occurred at 4° C but far more rapidly at 37° C. Higher binding in target cell chromatin (prostate as against spleen). Chromatin sites were saturable. Highly purified receptor complex still active in transfer.	MAINWARING and PETERKEN (1971) MAINWARING and IRVING (1973)
	(b) Unfractionated cytoplasm	Millipore membranes	Findings in accord with MAINWARING and PETERKEN (1971).	STEGGLES et al. (1971)

(3) Soluble extracts of nuclei in 0.5M KCl and purified DNA	Fractionated cytoplasm [b]	Formation of insoluble precipitate; extraction with 0.5M KCl and sucrose gradients	Complete dependence on complex II.[b] More precipitation with prostate than liver nuclear extracts.	TYMOCZKO and LIAO (1971)
(4) KCl-soluble nuclear extracts, DNA and synthetic polyribonucleotides	Fractionated cytoplasm	As in (3)	Complete dependence on complex II. Nuclear component a heat-labile, nonhistone protein. No specificity in the source of DNA. Poly(G) and poly(A) aided precipitation; poly(C) and poly(U) did not.	LIAO et al. (1972)
(5) Purified DNA alone; immobilised on cellulose [c]	Unfractionated cytoplasm or partially purified receptor	Proteins bound with a high affinity eluted in 0.5M KCl	Complete dependence on receptor protein; SBG was inactive. No clear evidence of saturation. No specificity in the source of DNA.	MAINWARING and IRVING (1973)
(6) Purified nuclei	Unfractionated cytoplasm or partially purified receptor	Extraction of [^{3}H] steroid into methylene chloride	Reactions occurred only at 37° C. Binding of testosterone and 5α-dihydrotestosterone without receptor. Significant retention only when receptor present. No assessment of relative transfer into nuclei of different sources.	NOZU and TAMAOKI (1975)

[a] only procedure used by MAINWARING (1970b); [b] FANG and LIAO (1969) distinguish two proteins, α and β, in prostate cytoplasm; both bind 5α-dihydrotestosterone, forming complexes I and II, respectively; [c] use of DNA-cellulose chromatography

The situation is less clear in rat prostate but three lines of evidence favour the temperature dependence of an obligatory change in the structure of the cytoplasmic receptor prior to its association with the nucleus. First, transition of the 8S receptor complex to a 3.5–4.5S form occurs before nuclear binding *in vivo* (MAINWARING and PETERKEN, 1971) yet the 3.5S complex II reacts readily with nuclear components even at 0° C (TYMOCZKO and LIAO, 1971). Second, the 8S receptor complex can react with chromatin at 4° C, but the interaction proceeds more rapidly at 37° C (MAINWARING and PETERKEN, 1971). A more rapid rate of dissociation of 5α-dihydrotestosterone from chromatin occurs at the higher temperature but this may be due to the thermal sensitivity of the acceptor components (TYMOCZKO and LIAO, 1971). Third, MAINWARING and IRVING (1973) showed that brief incubation of the 8S cytoplasmic receptor complex at 25° C converts it to a form indistinguishable in many respects from the active cytoplasmic complex II (FANG and LIAO, 1969) and the nuclear form of receptor extractable in 0.5 M KCl (MAINWARING, 1969 b). Most important of all, however, the activated receptor was rapidly bound to chromatin, *in vitro,* even at 4° C. This change in the quaternary structure of the cytoplasmic receptor complex will only be known with certainty when it has been purified to a form approaching homogeneity. Such investigations are not yet possible.

Receptor activation has been extensively documented in calf uterus (PUCA et al., 1972), rat kidney (MARVER et al., 1972), hepatoma cells (HIGGINS et al., 1973), and rat liver (KALIMI et al., 1975). In all these instances, receptor transformation may be simulated in cell-free conditions by careful manipulation of temperature and ionic strength, provided that the appropriate steroid hormone is present. All the activated complexes bind more rapidly but not more extensively to nuclei or chromatin. However, the transformed receptor need not necessarily have profoundly different physical properties other than the higher rate of association with chromatin. The cytoplasmic glucocorticoid receptor in rat liver, for example, may be activated in the cold by Ca^{2+} ions and without any change in sedimentation coefficient (KALIMI et al., 1975). The putative transforming factors of calf uterus (PUCA et al., 1972) have not been reported elsewhere.

g) The Structure of Chromatin and the Nature of Nuclear Acceptor Sites

Until recently, further understanding of the association of receptor complexes within the nucleus was hampered by the uncertainties surrounding the structure of chromatin. This stricture is important to studies on the location and nature of the acceptor components and in the long term, to investigations on the role of steroid-receptor protein complexes in metabolic regulation. For these reasons, present concepts of the structure of chromatin will be discussed.

Because of the extreme length of DNA relative to that of eukaryotic chromosomes, "packaging" or complex folding is needed to accommodate packing ratios in excess of 100:1 (DNA length:chromosome length; DU PRAW, 1970). As judged by x-ray diffraction, DNA in chromatin has a helical structure, but the more basic histones impose a state of supercoiling or packaging, thus reducing the length of chromatin strands (RICHARDS and PARDON, 1970). DNA-histone interactions are reciprocated in the sense that DNA also imposes conformational changes in the associated histones

(RAMM et al., 1972). Chromosome-associated RNA has been a contentious issue. While dismissed as a preparative artefact by several investigators (HEYDEN and ZACHAU, 1971; ARTMAN and ROTH, 1971), it is considered by others as playing an important role in the tissue-specific aspects of chromatin structure and function (MAYFIELD and BONNER, 1971; HOLMES et al., 1972). The distribution of the proteins in chromatin has long been the centre of widespread interest. The histones are certainly located in the major groove of DNA (WILKINS, 1956) and although probably repressors of genetic transcription, this histone function may be modified by methylation, phosphorylation, and acetylation (ALLFREY, 1971). The nonhistone proteins have been the focus of much attention and while considerable evidence supports the view that they impart the specificity to chromatin (for review, see PAUL, 1970), their location in chromatin was for long uncertain.

Over the last three years, concepts of the structure of chromatin have been revolutionised and a far more regular structure of repeating units is now envisaged. The contemporary picture of chromatin is a linear sequence of spherical chromatin particles, 7 nm in diameter, joined by long threads of only 1.5 nm in diameter, rather like beads along a string (KORNBERG, 1974). Evidence favouring this model comes from many lines of enquiry. First, WOODCOCK (1973) and OLINS and OLINS (1974) observed spherical particles by electron microscopy in specimens fixed and mounted in a special manner; the term υ bodies was coined for these particles by OLINS and OLINS (1974) whereas others have nominated them as nucleosomes (OUDET et al., 1975). Second, prompted by the work of HEWISH and BURGOYNE (1973) and RILL and VAN HOLDE (1973), digestion of chromatin with nucleases produces residual DNA subunits of 205 base pairs and 170 base pairs (NOLL, 1974), whereas endonuclease digestion produces a DNA subunit of 200 base pairs (KORNBERG, 1974; BURGOYNE et al., 1974). Third, it is now considered that histones are discrete oligomeric structures (KORNBERG and THOMAS, 1974; ROARK et al., 1974) and the fundamental repeating unit can be reconstituted with DNA and histones alone (AXEL et al., 1974; OUDET et al., 1975). The most persuasive model for chromatin consists of a flexible chain of nucleosomes, each of which contains equal weights of the four histones $F2a_1$, $F2a_2$, F2b, and F3 and DNA approximating to a length of 200 base pairs (OUDET et al., 1975; BALDWIN et al., 1975). This is diagramatically represented in Figure 5 from the studies by BALDWIN et al. (1975) on chromatin structure using neutron scattering techniques.

The concerted cross-linking effected by the four histones in nucleosomes is supported by trypsin-resistant cores described by WEINTRAUB (1975). The absence of histone F1 from nucleosomes should not obscure its biological importance. This histone is necessary for the packaging of the double helices of DNA in the strand-like regions (LITTAU et al., 1965; BRADBURY et al., 1974; VOGEL and SINGER, 1975). Furthermore, phosphorylation of histone F1 is believed to be a necessary prerequisite to chromosome condensation and cell division (BRADBURY et al., 1973; BRADBURY et al., 1974). The positions of the five major histones are now known with some precision and it is reasonable to conclude that nonhistones play little if any part in the assembly of the nucleosomes. In an important paper, LACY and AXEL (1975) isolated DNA from the repeating units of liver chromatin and compared the kinetics of the reassociation of subunit (nucleosome) DNA with total nuclear DNA. They established that all DNA sequences in the genome were present in the nucleosomes, in-

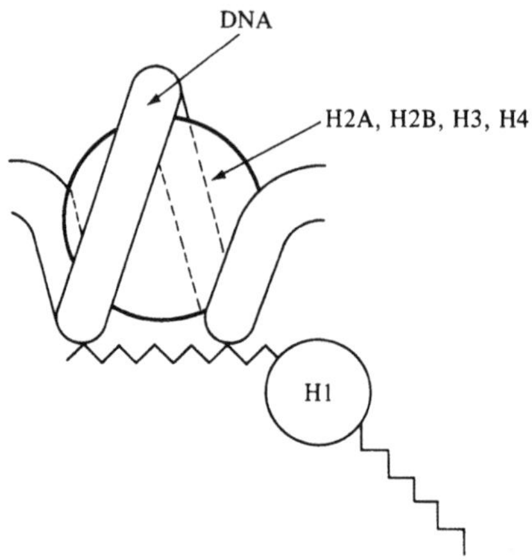

Fig. 5. A possible model for the subunit of chromatin. Protein core is composed of the apolar segments of four histones, with polar (basic) segments binding with DNA on outside of unit (equivalent to a ν body). Histone *H1* (equivalent to F1, in older nomenclature) is situated outside unit and may be involved in cross linking, either between subunits of same chromatin strand or between subunits of different strands. Figure is taken from BALDWIN et al. (1975) Nature Lond. 253, 245

dicating that the formation of nucleosomes is random with respect to base sequence. From additional studies, LACY and AXEL (1975) found that nucleosomes do not restrict the transcription of DNA and if indeed nonhistone proteins are genetic regulators, then they exert this control by recognising specific sites within the nucleosomes and the strands.

LANGMORE and WOOLLEY (1975) prepared chromatin under conditions where histone migration and proteolysis were minimised and examined the specimens without staining by using scanning electron microscopy. They suggest that chromatin consists of particles 13.4 nm in diameter, spaced about 26.2 nm apart along a narrow fibre. LANGMORE and WOOLLEY (1975) describe the particles as repeating units but these are considerably larger than the nucleosomes (7 nm diameter) described by others. Clearly "beads along a string" succinctly describes current ideas of chromatin structure, but the size of the repeating units remains in debate.

Against this background, a more positive approach can now be adopted in the interaction of receptor complexes with chromatin. Methods are also available for fractionating chromatin into active and inactive regions with respect to template activity for exogenous RNA polymerase. Fractionation procedures include differential centrifugation (FRENSTER et al., 1963), limited digestion with spleen deoxyribonuclease II at an alkaline pH far removed from its optimum (MARUSHIGE and BONNER, 1971) and chromatography on ECTHAM-cellulose (REECK et al., 1972). This cationic adsorbent is made by coupling tris(hydroxymethyl)aminomethane to cellulose with epichlorhydrin.

Using the procedure of MARUSHIGE and BONNER (1971), it has been possible to prepare active and inactive fractions from prostate chromatin (MAINWARING, 1975). These results are presented in Table 8 and justification of my nomenclature is evident from the experimental findings. Active chromatin had some very interesting prop-

Table 8. A comparison of the properties of fractions of prostate chromatin.

Prostate chromatin was separated by the procedure of MARUSHIGE and BONNER (1971) into active and inactive fractions. The results are taken from MAINWARING (1975) J. Reprod. Fest. 44, 375.

Fraction	Relative amounts[a] (% chromatin DNA)	Chemical composition (DNA = 1.0) Histone-protein	Nonhistone protein	Endogenous RNA polymerase [c] (c.p.m. $[^3H]$ CTP incorporated/250 μg chromatin DNA)	Acceptor activity [d] (c.p.m. $[^3H]$ 5α-dihydrotestosterone transferred/50 μg chromatin DNA)
Unfractionated chromatin	—	0.92 [b]	0.60 [b]	60 ± 7	1340 ± 20
Inactive chromatin	84–89	1.21–1.38	0.94–0.96	70 ± 11	810 ± 30
Active chromatin	11–16	0.11–0.18	1.26–1.38	610 ± 30	2100 ± 50

[a] based on DNA recovered, not initial DNA; [b] taken from MAINWARING and PETERKEN (1971); [c] the assay required the presence of all four nucleotide 5′-triphosphates; the enzyme is sensitive to α-amanitin yet stimulated by 0.2 M $(NH_4)_2SO_4$; [d] measured at saturating levels of prostate cytoplasmic receptor, labelled with $[^3H]$ 5α-dihydrotestosterone and corrected for the background transfer of free $[^3H]$ steroid.

erties, for not only was it conspicuously enriched with respect to nonhistone proteins but endogenous RNA polymerase and acceptor activities were preferentially concentrated in this fraction. At first sight, this appears to be a promising step forward, but two weaknesses are apparent. First, despite the usefulness of the fractionation procedure advocated by MARUSHIGE and BONNER (1971), it is difficult to envisage the location of the histone-depleted or active fraction in terms of the present concepts of chromatin structure. In a self-critical vein, it certainly would have improved this study if the histones and nonhistone proteins associated with the active and inactive fractions had been qualitatively analysed. It remains uncertain whether all or only a select few of the nonhistone proteins are retained in the active fractions. The omission of histone analysis was a more serious oversight, as histone F1 is perhaps the best marker for the strand regions of chromatin (see Fig. 5); such analyses would provide some clue to the location of the active chromatin fraction. Second, although acceptor activity is higher in active chromatin, this represents only a small proportion of total chromatin. Binding of receptor complex clearly occurs throughout chromatin and mainly in the inactive region on a proportional DNA basis.

In summary of receptor-chromatin interactions, it must be stated that current progress is somewhat disappointing, despite new insights into chromatin structure. Present methodology is incapable of distinguishing the specific acceptor sites for receptor complexes in chromatin. This point was cogently emphasized in a recent paper by YAMAMOTO and ALBERTS (1975), who drew heavily on recent experience with the specific binding of the *lac* repressor protein to operator regions of *Escherichia coli* DNA. These authors stressed that a small number of high affinity acceptor sites would not be detected by conventional techniques against a heavy background of low-affinity or nonspecific acceptor sites. On current estimates (see KING and MAINWARING, 1974), the number of steroid molecules entering the nucleus of a target cell is in the region of 5000–10,000; certain studies put it as high as 60,000 (BRUCHOVSKY et al., 1975). However, YAMAMOTO and ALBERTS (1975) suggest that the number of specific, high-affinity acceptor sites may be as low as 1000 and hence the need for sophisticated methods of analysis. This is extremely relevant to the results presented in Table 8. YAMAMOTO and ALBERTS (1975) proposed two new approaches to circumvent present problems. First, analyses should be based on the putatively longer dissociation time for receptor complexes from the specific acceptor sites in chromatin. This may be exploited in assays based on the retention of chromatin or DNA on membranes (for example, see HINKLE and CHAMBERLIN, 1972). Ideally, the binding component should be homogeneous and this is not yet possible with androgen receptor complexes. Second, the possibility of enriching the hormone-specific regions of the genome should be pursued. The exciting suggestion is to utilise a virus whose function or growth is ultimately under the control of a steroid-receptor protein complex; there is a precedent for such an interrelationship in the growth of polyoma virus in mouse 3T3 cells in the presence of dexamethasone (MORHENN et al., 1973).

The remainder of this section will be devoted to two current problems; the nature of the acceptor components and the part played by DNA in nuclear acceptor activity. To ease the problems of presentation, the involvement of DNA in acceptor activity will be discussed first.

Beginning with work from this laboratory (MAINWARING and MANGAN, 1971) it is now generally acknowledged that receptor complexes for all the major classes of steroid hormones possess the ability to bind to DNA (CLEMENS and KLEINSMITH, 1972; KING and GORDON, 1972; TOFT, 1972; HIGGINS et al., 1973 b; YAMAMOTO and ALBERTS, 1974). The methodology has included DNA-cellulose chromatography (MAINWARING and MANGAN, 1971; CLEMENS and KLEINSMITH, 1972), retention of DNA on glass fibre discs (KING and GORDON, 1972), sucrose gradients (TOFT, 1972; ANDRÉ and ROCHEFORT, 1973), and sedimentation partition chromatography (YAMAMOTO and ALBERTS, 1974). The most recent method for studying DNA-receptor complex interactions, suggested by ALBERGA et al. (1976), is based on the distribution of macromolecules between two aqueous phases containing the polymers polyethylene-glycol 6000 and dextran T500 (ALBERTSSON, 1965). DNA partitions solely in the dextran phase and this procedure has much to recommend it in terms of speed and flexibility.

The importance of the association of receptor complexes with DNA stems from the accepted dogma that hormones are ultimately bound within the nucleus and elicit their metabolic control by processes requiring DNA as template, including RNA polymerase and DNA polymerase (MAINWARING and MANGAN, 1971). Metabolic regulation by means of the interactions of proteins with DNA is well documented in microorganisms. Suitable examples include the blocking of the *lac* operon by its repressor (GILBERT and MÜLLER-HILL, 1967; RIGGS et al., 1970), the termination of RNA synthesis by ρ factor (ROBERTS, 1969), and the activation of DNA as a template for DNA synthesis by the gene 32 protein (ALBERTS and FREY, 1970; HUBERMAN et al., 1971). Taking the example of the *lac* repressor a little further, binding to DNA includes a limited number of specific, high affinity sites (GILBERT and MÜLLER-HILL, 1967) and a virtually unlimited number of nonspecific, low affinity sites (LIN and RIGGS, 1972). These observations are particularly relevant in the context of specificity of DNA-receptor complex interactions.

Despite their appeal, DNA-receptor complex interactions fail to demonstrate biological specificity. Derogations lie in the following considerations. First, with the notable exception of the study by CLEMENS and KLEINSMITH (1972), there is no specificity in the source of DNA concerning receptor binding; in all other cases, equal binding was reported to DNA isolated from eukaryotic and prokaryotic sources. Second, saturation of the binding sites in DNA has never been demonstrated unequivocally even at a high input of receptor complex (ANDRÉ and ROCHEFORT, 1973; ALBERGA et al., 1976). This indicates that much of the binding is of low affinity and hence of minor biological significance. This is in keeping with the views of YAMAMOTO and ALBERTS (1975) that all procedures lack the specificity to distinguish the specific sites in DNA needed for the expression of hormonal responses. With the *lac* repressor, verification of the small number of specific sites was available from mutants containing aberrations in the sequence of DNA adjacent to the *lac* operon (VON HIPPEL et al., 1974).

Before DNA-receptor complex interactions are dismissed out of hand, however, cognisance must be given to observations favouring their biological importance. Binding of steroid hormones to DNA cannot be mimicked by plasma proteins, despite the fact that they have a high affinity for steroid hormones (MAINWARING and IRVING, 1973; BULLOCK et al., 1975). The elegant studies by YAMAMOTO et al. (1974) strongly uphold the significance of DNA-receptor complex interactions. Clones were selected from mouse S49.1A lymphoma cells on the basis of their resistance or sensitivity to dexamethasone. YAMAMOTO et al. (1974) described differences in the nuclear mechanism for glucocorticoids in the variant clones, either as deficient (nt^-) or increased (nt^i) relative to the parental lymphoma cells. Both these classes of clones contained cytoplasmic glucocorticoid receptors, but the receptors from nt^- variants bound with a lower affinity to lymphoma DNA than their counterparts in nt^i variants. Together with differences in sedimentation behaviour, the contrasting ability of nt^- and nt^i receptors to bind to DNA provides the best explanations of the responses of the variant clones to dexamethasone.

The significance of the binding of steroid-receptor protein complexes to DNA cannot be critically appraised at the present time. Two schools of thought exist. Proponents of the active theory (SPELSBERG et al., 1971, 1972) attribute all acceptor activity to nuclear proteins with little if any involvement of DNA. Supporters of the passive theory (MAINWARING and PETERKEN, 1971; KING and GORDON, 1972) suggest that DNA is probably the acceptor while nuclear proteins restrict the binding sites available on DNA. These theories will be evaluated at the end of this section.

Past attempts to characterise the acceptor components have depended on rather indirect experimentation. Certainly chromosomal RNA is not involved as pretreatment of chromatin with electrophoretically pure ribonuclease which did not impair retention of the cytoplasmic androgen receptor complex (MAINWARING and PETERKEN, 1971). Other evidence implicated nonhistone proteins in the acceptor function of chromatin (TYMOCZKO and LIAO, 1971; MAINWARING and PETERKEN, 1971; SPELSBERG et al., 1971, 1972). However, it must be emphasized that this inference was drawn from the finding that acceptor activity resided in an AP_3 or nonhistone protein fraction, extractable only in 5 M urea at pH 8.3 (SPELSBERG et al., 1971, 1972), but the acceptor component was never characterized.

About two years ago, the limitations of the methods used for studying acceptor-receptor complex interactions were becoming very apparent. Irrespective of the form of the acceptor components, techniques suffered from experimental variability and tedious washing procedures to eliminate unbound or excess receptor complex. The failure to demonstrate the saturation of acceptor sites was rightly stressed by CHAMNESS (1973) and artefactual binding of receptor aggregates was emphasized by YAMAMOTO (1976). Many of these problems were adroitly circumvented by PUCA et al. (1974), whose method was elegant in its simplicity. Acceptor components were extracted from purified nuclei in 2 M NaCl and covalently linked to Sepharose 4B, thus forming a matrix suitable for affinity chromatography. Receptor complex, labelled with the appropriate [^{3}H] steroid ligand, is passed through the matrix and after extensive washing, the receptor complex bound with a relatively high affinity can be eluted in media of high ionic strength. The advantages of this approach are threefold; bulk preparation of a stable acceptor matrix, extreme analytical precision, and experimental flexibility.

The procedure of PUCA et al. (1974) has recently been adopted in this laboratory and up to 60 columns may be processed concomitantly if the matrix is packed in disposable Pasteur pipettes (MAINWARING et al., 1976 b). The principal features of the interactions between immobilised acceptor sites and the cytoplasmic androgen receptor complex of rat prostate are as follows. (1) Under the conditions of analysis employed, retention of the receptor complex was completely dependent on the presence of immobilised acceptor sites; no retention was found to Sepharose 2B alone, provided that the activated groups introduced by CNBr (AXÉN et al., 1967) were blocked by glycine. Ethanolamine is probably a more effective blocking agent (HEINZEL et al., 1976). (2) The receptor complex must be in the native or biologically active configuration because the modification of essential -SH groups with 1 mM N-ethylmaleimide (MAINWARING, 1969 a) negated binding to the affinity matrix. (3) The receptor protein was retained only when presented to the acceptor sites in the presence of its favoured ligand, [^{3}H] 5α-dihydrotestosterone. If steroid-free cytoplasm was washed through the column first and [^{3}H] 5α-dihydrotestosterone was later applied alone, there was no retention of radioactivity. However, if the two wash fractions of [^{3}H] 5α-dihydrotestosterone and steroid-free cytoplasm were mixed together and reapplied to the matrix, then significant retention of radioactivity occurred. (4) When increasing amounts of receptor complex were applied to a fixed amount of matrix, acceptor saturation was found. Two sets of acceptor sites were identified; a small number of saturable, high-affinity sites and a vast number of seemingly unsaturable, low affinity sites. The apparent K_d of the high affinity sites was 2.5×10^{-9} M. (5) The two sets of acceptor sites were distinguished by using 0.5 M KCl, the salt concentration required to release receptor complex from intact nuclei (MAINWARING, 1969 b). The low affinity could be occupied in 0.5 M KCl whereas the high affinity sites could not (see Fig. 6.). (6) The affinity matrix from prostate nuclear extracts contained 470 μg protein, 40 μg DNA and 40 μg RNA covalently linked to 1 g wet weight of Sepharose 2B. From the effects of enzymes of limited substrate specificity, the nucleic acids played no part in the high affinity retention of the receptor complex; the small number of specific sites were present, therefore, in nuclear proteins. Our findings were totally substantiated by more recent work by PUCA et al. (1975 a, b);

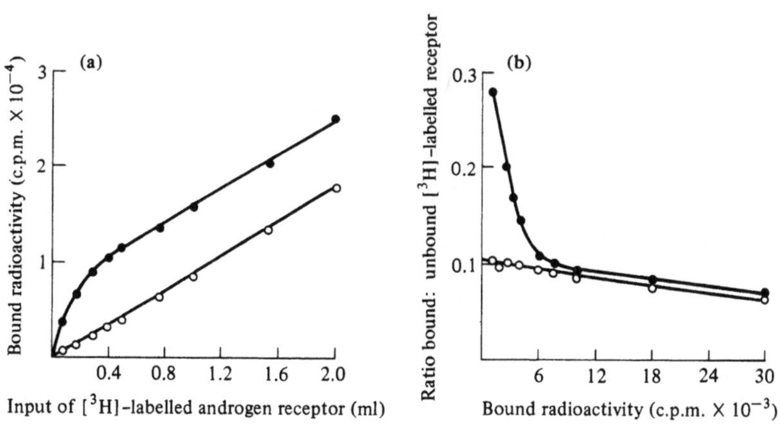

Fig. 6 a and b. The binding properties of immobilised acceptor sites of rat prostate. Nuclear extracts were immobilised on Sepharose 2B and cytoplasmic receptor, labelled with 0.5×10^{-9} M [^{3}H] 5α-dihydrotestosterone, were selectively precipitated in 33% saturated $(NH_4)_2SO_4$. Columns containing 500 mg wet weight of affinity matrix were treated with increasing amounts of [^{3}H]-labeled receptor complex. Interactions in absence, ●, or presence, ○, of 0.5 M KCl. (a) Conventional saturation analysis; (b) Data plotted according to SCATCHARD (1949). Results are taken from MAINWARING et al. (1976) Biochem. J. 156, 253 the oestrogen receptor in calf uterus is bound to its acceptor sites more avidly (K_d 2×10^{-10} M).

A more detailed examination of androgen receptor-acceptor interactions is presented in Table 9. There was an absolute requirement for the androgen receptor protein. Free [^{3}H] 5α-dihydrotestosterone was not retained and the role of the receptor protein could not be mimicked by soluble proteins of nontarget cells or the plasma protein, SBG. The mandatory requirement for [^{3}H] 5α-dihydrotestosterone was es-

Table 9. Specificity of receptor complex-acceptor interactions; use of affinity chromatography.

Nuclear proteins, solubilised in 2M NaCl were coupled to CNBr-activated Sepharose 2B. Receptor preparations in rat prostate 100,000 *g* supernatant fractions were labelled with 5×10^{-9}M[^{3}H] 5α-dihydrotestosterone. Labelled samples were applied to affinity chromatography matrix and bound proteins were eluted in 1M NaCl. Data are taken from Mainwaring et al. (1976) Biochem. J. **156**, 129.

Nature of experiment	Particular conditions	Bound [^{3}H] 5α-dihydrotestosterone [a] (c.p.m./500 μg of immobilised protein)
(a) Specificity in androgen receptor protein (Prostate acceptor matrix)	Source of cytoplasm [b]	
	Prostate	4140
	Prostate; 33% sat. $(NH_4)_2SO_4$	3720
	Prostate; 33–66% sat. $(NH_4)_2SO_4$	320
	Liver	220
	Kidney	150
	Spleen	120
	Serum (human) [c]	220
	None; free [^{3}H] 5α-dihydrotestosterone	110
(b) Specificity in ligand, 5α-dihydrotestosterone (Prostate acceptor matrix and prostate cytoplasm)	^{3}H 5α-dihydrotestosterone with 20-fold excess non-radioactive competitor	
	No competitor	4490
	5α-Dihydrotestosterone	690
	5β-Dihydrotestosterone	4400
	Dexamethasone	4670
	BOMT (antiandrogen)	
(c) Distribution of acceptor activity. (Prostate cytoplasm)	Source of nuclear extract [d]	
	Prostate	4600
	Liver	650
	Kidney	1470
	Spleen	2030
	Pancreas	410
	Lung	1170

[a] experimental error, ± 5% only; [b] all at 8 mg protein/ml; [c] source of SBG; [d] diluted with uncoupled Sepharose 2B to common amount of immobilised protein (500 μg/500 mg wet weight).

tablished from experiments where the receptor protein was labelled in the presence of nonradioactive steroid competitors. Glucocorticoids and 5β-reduced steroids have no influence on cytoplasmic receptor binding (MANGAN and MAINWARING, 1972) and did not impair receptor-acceptor interactions. By contrast, almost complete competition was achieved with nonradioactive 5α-dihydrotestosterone and also by the antiandrogen, BOMT (see MANGAN and MAINWARING, 1972). When affinity matrices from different sources were compared, the androgen receptor complex bound more extensively to prostate nuclear proteins. This strengthens the credibility of the acceptor hypothesis because the present differences cannot be easily explained by changes in the structure of chromatin during its isolation or to other flaws in experimental design. The specificity reported here is again supported by the recent work of PUCA et al. (1975 b) on oestrogen-sensitive tissues.

Affinity chromatography has proved to be of great value for studying receptor complex-acceptor interactions, and under our conditions of analyses, DNA is not involved. In the rat prostate nucleus, there seem to be about 500 to 1000 protein acceptor sites capable of binding the cytoplasmic receptor complex with a high affinity but a vast excess of nonspecific acceptor sites are also present. The principal determinants in the specificity of this nuclear binding process reside in cytoplasmic elements, especially in the receptor protein itself and its optimum natural ligand, 5α-dihydrotestosterone. A certain measure of specificity is also imparted by the acceptor protein moiety. Future work must also include studies on the activated form of the cytoplasmic receptor complex.

Until the work of PUCA et al. (1974, 1975 a) was published, acceptor proteins were said to be nonhistone proteins with presumably a negative overall charge. However, PUCA et al. (1975 a) suggested that the acceptor activity in calf uterus resided in a basic nuclear protein; evidence from this laboratory supports the basic nature of the high affinity acceptor components in rat prostate (MAINWARING et al., 1976 b). (1) When nuclear proteins were fractionated by ion-exchange chromatography on Bio-Rex 70 columns (LEVY et al., 1972; RICHTER and SEKERIS, 1972), acceptor activity was concentrated in the basic protein fraction. (2) On isoelectric focusing in a gradient of 10–60% sucrose in 3 M urea, the high affinity acceptor components were found to have their pI in the range 8.6–9.3; the low affinity acceptor compounds were recovered at pH 7.0 and below. The only disquieting aspect of this analysis was the high concentration of urea required to keep the acceptor proteins in solution under conditions of low ionic strength; a certain measure of protein denaturation cannot be discounted. (3) In calibrated columns of Sephadex G-75, the acceptor proteins had a mean molecular weight of 70,000. (4) The high affinity acceptor proteins had a sedimentation coefficient of approximately 6.5S in sucrose gradients run in the presence of 2 M NaCl. By electrophoretic analyses, the classical histones could not be detected in the area of the gradient profile containing acceptor activity. Despite their basic charge, the high affinity acceptor components do not appear to be histones on the basis of these physicochemical properties (see BRADBURY and CRANE-ROBINSON, 1971; HNILICA, 1972). It is also difficult to envisage how the acceptor proteins could be histone aggregates, because our analyses were conducted either in the presence of 2 M NaCl or 3 M urea. The basic pI of the acceptor protein fractions runs counter to the acidic character of most nonhistone proteins; they are best described as nonhistone proteins with a basic overall charge. Our evidence does not enable one to say whether the high affinity acceptor function is carried out by a single protein or a family of closely related proteins. This question can only be answered when the acceptor activity has been extensively purified.

Since these affinity chromatography studies do not implicate DNA, their relevance to DNA-receptor complex interactions should be broached. With regard to the earlier studies by Spelsberg et al. (1971, 1972), it is highly probable that small amounts of a basic nonhistone protein, such as the acceptor, would escape detection by present methods of electrophoretic analysis of their AP_3 fraction; this predominantly contained acidic nonhistone proteins. Furthermore, the absolute resolution of acidic and basic nonhistone proteins by their relative solubility in buffered 5 M urea has still to be established. At first glance, the present studies tend to favour the active theory of DNA involvement in acceptor activity. If, however, the basic acceptor proteins had a high propensity for DNA, then current differences of opinion between the active and passive theories of DNA and acceptor activity would be largely reconciled.

h) The Exit of 5α-dihydrotestosterone from the Prostate Gland: Receptor Cycling

The high-affinity binding mechanism for 5α-dihydrotestosterone in the nucleus of the prostate gland retains the active metabolite for a period of between 12 and 16 hours (TVETER and ATTRAMADAL, 1968; MAINWARING and PETERKEN, 1971). Very little is known of the mechanism responsible for the exit of 5α-dihydrotestosterone and the receptor protein from the nucleus. To a certain extent this impasse is understandable, because it is difficult to devise experiments in the intact cell to study this problem. It should also be added, however, that investigators in general are far more interested in establishing how hormonal mechanisms are "switched on" or activated, rather than how they are "turned down." In the long term, this prejudice could hamper the advancement of knowledge of hormonal mechanisms, particularly with respect to developmental changes. As discussed in Chapter II, development plays a significant part in the mechanism of androgen action.

Although the metabolism of testosterone to 5α-dihydrotestosterone is a critical early event in the androgenic response in rat prostate, there is no clear evidence that further metabolism of the active metabolite plays a part in its exit from androgen target cells. Furthermore, the work of BELL and MUNCK (1973) and CASTEÑADA and LIAO (1975) suggests that the envelopment of steroids within receptor proteins reduces their availability to other external proteins; this is certainly true for immunoglobulins (CASTEÑADA and LIAO, 1975) and perhaps for steroid-catabolising enzymes. However, this view should be accepted with a certain measure of caution in the light of recent studies by NOZU and TAMAOKI (1975 b). These authors incubated the 9S cytoplasmic receptor protein- [^{3}H] 5α-dihydrotestosterone complex with 3α-hydroxysteroid dehydrogenase and observed a rapid conversion of the bound active metabolite to 5α-androstane-3α, 17β-diol. This androgen-related steroid has negligible affinity for the receptor protein and these interesting experiments suggest a possible mechanism for the degradation of androgen receptor complexes in the intact cell. However, the importance of the work of NOZU and TAMAOKI (1975 b) would be considerably enhanced if they could demonstrate a similar breakdown of the nuclear-bound 5α-dihydrotestosterone; one would imagine that the active metabolite retained in chromatin is far less accessible than that bound to the cytoplasmic receptor alone.

Glucocorticoid target cells provide more favourable systems for studies on the entry and exit of steroids. Clear evidence for the active transport of glucocorticoids into pituitary adenocarcinoma cells maintained in culture has been presented by HARRISON et al. (1975). A specific exit mechanism has also been reported in dexamethasone-sensitive sublines of mouse L929 fibroblasts (GROSS et al., 1968, 1970). Such a system for the active transport of steroids out of cells has not been demonstrated unequivocally in other systems and certainly not in androgen target cells. By careful control of temperature and other factors, the binding of glucocorticoids in lymphocytes can be stopped at selected stages of the overall binding process. By such techniques, MUNCK and BRINCK-JOHNSEN (1974) specifically studied the release of [^{3}H] cortisol from lymphocyte nuclei. Although exceedingly sensitive to steroid-catabolising enzymes, MUNCK and BRINCK-JOHNSEN (1974) could not detect any metabolic breakdown of [^{3}H] cortisol during its release from the nucleus. In summary, little is known of the processes by which steroid hormones leave chromatin or the cell.

LIAO et al. (1973 b, 1973 c) have propounded the concept that the androgen receptor protein is cycled between the cytoplasm and nucleus of rat prostate. After initial binding of 5α-dihydrotestosterone to the receptor protein in the cytoplasm, the resultant complex is then translocated to the acceptor sites in chromatin. The principal idea of receptor cycling is that the receptor complex may then bind to nuclear particles rich in ribonucleoproteins and be transported back into the cytoplasm where the carrier ribonucleoprotein particles may participate in protein synthesis (LIAO et al., 1973 b, c). By recycling, the receptor may be engaged in the processing and distribution of nuclear species of RNA which could otherwise remain within the nucleus. This is a very plausible concept, but like other possible mechanisms, it still does not explain how 5α-dihydrotestosterone finally leaves the prostate cell. More work is needed in this area of the mechanism of action of androgens.

i) Binding of Steroids Other than 5α-dihydrotestosterone

Binding of progesterone (KARSZIA et al., 1969) and oestradiol-17β (UNHJEM, 1970 b) has been reported in rat prostate cytoplasm. It is difficult to attribute biological significance to these proteins for two reasons: these steroids do not bind to prostate nuclei to any appreciable extent and they do not induce macromolecular synthesis in male accessory sexual glands.

Following in the wake of earlier studies by UNHJEM and TVETER (1969), the nuclear binding of testosterone in rat prostate has been widely documented (RENNIE and BRUCHOVSKY, 1972; BRUCHOVSKY et al., 1975; NOZU and TAMAOKI, 1975 a). Using chromatography on phosphocellulose, RENNIE and BRUCHOVSKY (1972) resolved two forms of receptor in rat prostate, one in the cytoplasm and one in the nucleus; they also reported that nuclear binding of 5α-dihydrotestosterone declined after castration whereas the nuclear retention of testosterone was not affected. BRUCHOVSKY et al. (1975) reported that up to 60,000 molecules of androgenic steroid may be retained within the nucleus of a single prostate cell, but during this process, the number of receptor molecules depleted in the cytoplasm or acquired within the nucleus is not commensurate with the large number of steroid molecules transported. While it remains possible that many steroid molecules may be transported by a single molecule of receptor protein, BRUCHOVSKY et al. (1975) suggest that the concentration of androgens in the prostate nucleus is only partially regulated by the cytoplasmic receptor protein and that androgens can enter chromatin by means independent of the receptor system. The study by NOZU and TAMAOKI (1975 a) supports this view. They reported that both testosterone and 5α-dihydrotestosterone could enter purified prostate nuclei at 37° C in the absence of cytoplasmic proteins; however, cytoplasmic receptor protein facilitated the retention of only 5α-dihydrotestosterone within the nuclei. The survey of reconstituted, cell-free systems for the transfer of 5α-dihydrotestosterone into prostate chromatin or nuclei (section III.1.f.) and particularly the affinity chromatography work (section III.1.g), attributes crucial importance to the cytoplasmic androgen receptor protein in the association of 5α-dihydrotestosterone with the acceptor components. In all cell-free systems, testosterone could not replace 5α-dihydrotestosterone. On current evidence, therefore, these biologically active androgens are transported into prostate nuclei by sharply contrasted mechanisms; transport of 5α-dihydrotestosterone has an obligatory requirement for the receptor protein but testosterone may be retained in the nucleus by receptor-independent mechanisms.

If this conclusion is essentially correct, then three outstanding issues remain. First, the metabolic effects produced by the direct attachment of testosterone to nuclei or chromatin warrant extensive exploration. Second, the metabolic control exerted by receptor protein-5α-dihydrotestosterone complexes can be tested experimentally, albeit in a rather unsophisticated manner (see section III.4.2.); it is essential that effects of testosterone should be strictly monitored under these conditions. These experiments would provide some indication of the importance of the binding of testosterone to the prostate nucleus. Third, the relative importance of 5α-dihydrotestosterone and the receptor protein should be considered. Throughout this text and many papers in the literature, the function of the receptor protein is seen as a sophisticated transport vehicle for a specific, active metabolite, namely 5α-dihydrotestosterone. In terms of biological importance, the reverse is probably true in that 5α-dihydrotestosterone enables an important regulatory protein to gain access to the genetic apparatus; in essence, the two constituents of the receptor complex are equally important. Attachment of a specific protein to restricted portions of the genome is an efficient way of initiating a precise sequence of biochemical events. Many such instances may be cited in microorganisms (GILBERT and MÜLLER-HILL, 1967; ROBERTS, 1969; ALBERTS and FREY, 1970; VON HIPPEL et al., 1974). The formation of the receptor complex transforms a "low information" steroid molecule into a "high information" steroid protein complex. The high affinity of the interaction results in economy of steroid molecules by protecting them from enzymic breakdown or indiscriminate attachment to macromolecules of lower binding affinity.

j) Purification of Receptor Proteins

The purification of androgen receptor complexes is of extreme importance, not only for fundamental studies on receptor site-ligand interactions, but also for investigating the importance of the receptor proteins as metabolic regulators, especially of genetic transcription. Receptor purification is a daunting task, because the receptor proteins are notoriously labile (MAINWARING, 1969 a; BAULIEU and JUNG, 1970; FANG et al., 1969) and they are present in only minute quantities in androgen target cells. MAINWARING and IRVING (1973) made a conservative estimate that 40 μg of receptor protein is present in a kilogram of rat prostate tissue or expressed another way, the ventral prostates of 4000 rats. Another serious drawback is that the only reliable means for detecting the receptor protein is its association with a tritiated ligand of high specific radioactivity. Since the nuclear receptor complex may only be formed in intact cells or in reconstituted, cell-free systems *in vitro,* the difficulty of labelling it extensively renders its purification even more difficult. On the other hand, the cytoplasmic receptor may be labelled directly in the cold and the ensuing presentation is directed solely at the purification of the soluble (cytoplasmic) androgen receptor complex.

It is clear that conventional methods of protein fractionation will be pushed to the limits of their resolution in the purification of receptor complexes. The best effort thus far is the 5000-fold purification of the 8S form of the cytoplasmic receptor complex achieved by MAINWARING and IRVING (1973). The characteristics of the receptor complex exploited in this study were its ability to bind to DNA-cellulose under conditions where plasma proteins were not retained and its relatively acidic pI (5.8). Extensive dissociation of [^{3}H] 5α-dihydrotestosterone from the receptor complex was partially offset by the inclusion of the [^{3}H] ligand in all preparative media. The partially purified receptor complex fully retained its ability to bind to the acceptor sites of prostate chromatin. However, the limitations of this approach were exposed by estimates of the purity of the final product; at best, the receptor complex was only 5–8% pure. DNA-cellulose chromatography has also been used for the partial purification of the glucocorticoid receptor complex in rat liver (EISEN and GLINSMANN, 1975).

From this it follows that more sophisticated approaches must be adopted. In particular, dissociation of the bound [^{3}H] ligand must be prevented and the fractionation methods must be both rapid and specific to provide more stabilisation of the labile complex. Affinity labelling has been profitably used in the characterisation of the catalytic sites of steroid-metabolising enzymes. Structural analogues of the steroid substrate are selected which link covalently to the amino acid residues of the active site, thus enabling the active portion of the protein to be identified unequivocally. Warren and his co-workers have pioneered this approach and potential derivatives include 2-diazosteroids (CHIN and WARREN, 1970) and halogenated steroids, especially at the C-6 position (GANGULY and WARREN, 1971; ARIAS et al., 1973). In principle, affinity labelling should be applicable to the identification of any high-affinity binding site for steroids, including those in receptor proteins. Bearing in mind the importance of C6-halogenated derivatives in affinity labelling (ARIAS et al., 1973), analogues of cyproterone or BOMT could be useful candidates for identifying the binding sites in receptor proteins. While both antiandrogens have a relatively low affinity for the receptor binding site (see section III.1.d), this disadvantage will be outweighed by their even lower affinity for the androgen-binding proteins of plasma (see Table 5.) and their resistance to steroid catabolism. The most striking application of the affinity labelling of receptor sites is in the study by KATZENELLENBOGEN et al. (1974). They labelled the oestrogen receptor of calf uterus with light-sensitive analogues of oestradiol which were covalently attached to the receptor site after ultraviolet irradiation. Similar principles were employed by WOLFF et al. (1975) in the affinity labelling of the glucocorticoid receptors in rat kidney. Perhaps the only pitfall to be avoided in this procedure is the unwitting attachment of the light-sensitive ligands to low-affinity binding proteins, especially those derived from plasma. Should this occur, then precise methods must be used to distinguish the affinity label attached to the receptor proteins.

At the present time, only the procedure of affinity chromatography seems to offer the necessary specificity for the extensive purification of androgen receptor complexes, but no reports are currently available. Nevertheless, impressive advances have been made in the purification of the receptors for other steroid hormones using affinity chromatography, including oestradiol-17β (SICA et al., 1973) and especially progesterone (SMITH et al., 1975; KUHN et al., 1975). While a steroid-containing affinity matrix has not yet been developed specifically for the isolation of androgen receptor proteins, several have been proposed for the purification of human SBG with various measures of success (BURSTEIN, 1969; MICHELSON and PÉTRA, 1975; ROSNER and SMITH, 1975). There is no a priori reason why matrices containing immobilised 5α-dihydrotestosterone should not be equally applicable to receptor purification; a survey of possible candidates for affinity chromatography is presented in Figure 7. All these matrices contain 5α-dihydrotestosterone coupled by spacer groups at either the C-3 or C-17 position to the polysaccharide, agarose, available commercially as Sepharose 2 or 4B. One could predict from first principles that a matrix with the mandatory 17β-hydroxyl group in a free (unsubstituted) form would be more efficient (see section III.1.c.) and this is indeed the case. The matrix devised by ROSNER and SMITH (1975) is superior to that of MICKELSON and PÉTRA (1975); SBG is purified on the matrix substituted at C-3 (free 17β-hydroxyl group) with a greater facility and higher purity. The selection of the spacer group is still based on empirical judgments, but it is a critical feature of affinity chromatography. To emphasize this point, the matrix suggested by BURSTEIN (1969) has no real spacer group yet it binds SBG so avidly that it can only be eluted under conditions of protein denaturation, for example using 4 M guanadinium hydrochloride. The matrix of choice seems to be that developed by ROSNER and SMITH (1975). Androgen-free prostate cytoplasm would be applied to the matrix and the absorbed receptor-protein subsequently eluted with 5α-dihydrotestosterone. High affinity, androgen-binding proteins from plasma would also be purified under these conditions and means must be found to specifically distinguish the androgen receptor complex. Three possibilities present themselves. (1) Differences in physical properties may be exploited, because highly purified SBG and CBG have lower pI's than androgen receptor complexes (MUIDOON and WESTPHAL, 1967; VAN BAELEN et al., 1972; ROSNER and SMITH, 1975). (2) Antibodies raised in rabbits against testosterone or 5α-dihydrotestosterone may be covalently linked to Sepharose 4B and this affinity matrix should selectively remove the SBG-5α-dihydrotestosterone complex. The receptor complex would not be retained as the steroid is enveloped and not accessible to the antibody (CASTEÑADA and LIAO, 1975). (3) The affinity matrix containing immobilised acceptor protein could be extremely efficient in selectively retaining the androgen receptor complex whereas SBG should pass through the column unimpeded (MAINWARING et al., 1976). It is hoped that methods such as these will be applied to the purification of androgen receptor complexes in the near future.

There seems little point in trying to label the receptor protein moiety of the complex with 5α-dihydrotestosterone as a means of identification during purification.

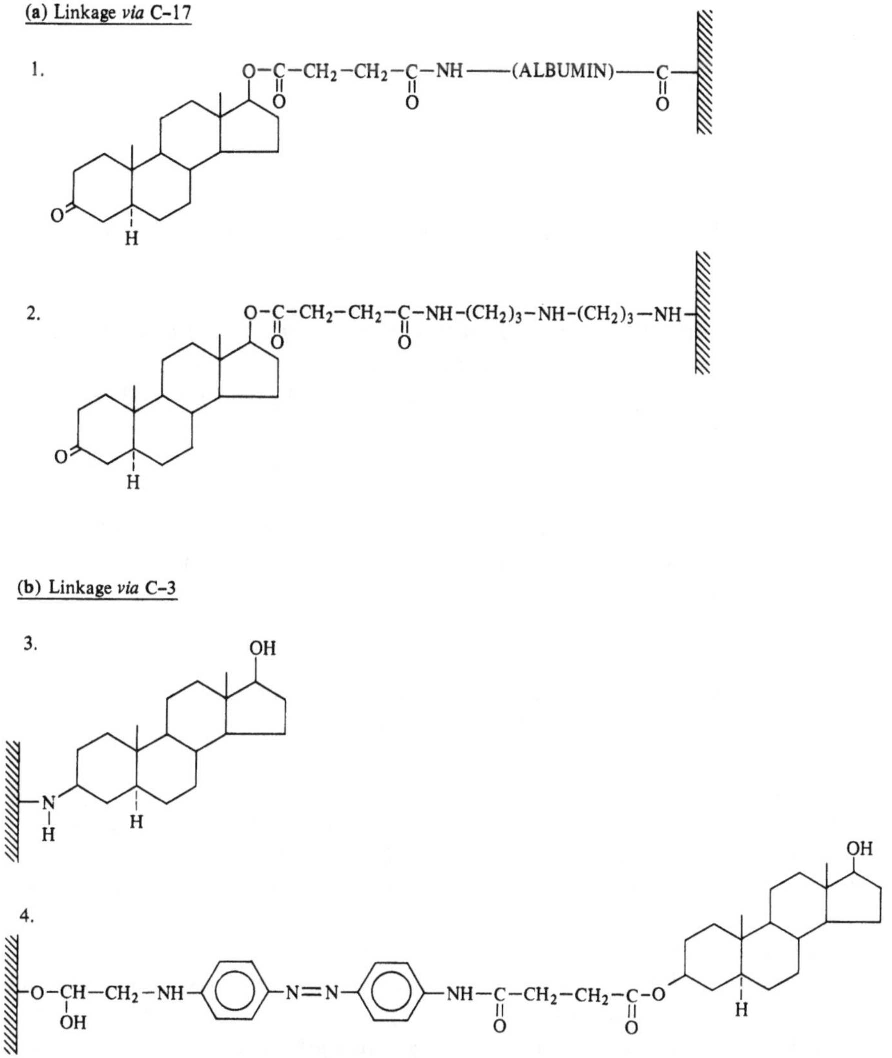

Fig. 7 a and b. Affinity matrices for the purification of androgen receptor protein complexes. Matrices contain derivatives of 5α-dihydrotestosterone immobilised by linkage via either C-17 (a) or C-3 (b). Derivatives: 1, 5α-dihydrotestosterone hemisuccinate coupled to Sepharose-albumin (MICKELSON and PÉTRA, 1975); 2, 5α-dihydrotestosterone hemisuccinate coupled to Sepharose-3,3′diaminodipropylamine (MICKELSON and PÉTRA, 1975); 3, 3β-amino derivative of 5α-dihydrotestosterone coupled to Sepharose (BURSTEIN, 1969); 4, 5α-dihydrotestosterone-3β-hemisuccinate coupled to Sepharose-azodianiline (ROSNER and SMITH, 1975)

However, NOZU and TAMAOKI (1974) labelled prostate cytoplasmic proteins with ^{131}I and claimed that the radioactively-labelled receptor complex could be identified after incubation with purified prostate nuclei. While admiring the enterprise of these investigators, the results must be considered debatable. The receptor protein constitutes such a minute fraction of prostate cytoplasmic proteins (MAINWARING and IRVING, 1973) that the nonspecific transfer of other iodinated proteins into nuclei would certainly obscure the transfer of ^{131}I attributable to the translocation of the receptor protein.

2. The Binding of Androgens in Other Target Cells

Androgens control the growth and function of so many different types of cells that a unified theme of androgen-binding mechanisms is difficult to present. The situation is complicated further by the fact that different target cells have received varying degrees of attention; in some instances, binding studies are advanced while knowledge of other systems is fragmentary. For the basis of discussion, it is most likely that the salient features of the binding mechanisms described in rat prostate are applicable to most androgen target cells. The major differences from one tissue to another are reflected more in their metabolic capabilities than in the presence or absence of receptor mechanisms.

a) Distribution and Specificity of Receptors for Androgens and Related Steroids

A restricted survey of androgen-binding mechanisms in various cells is presented in Table 10. In preparing this survey, particular attention was given to the ability of cells to metabolise testosterone and to bind selective metabolites; these are the key features of the mechanism of androgen action. The binding mechanism in the various androgen target cells will now be discussed in more detail.

As far as the male accessory glands are concerned, the binding mechanism described for 5α-dihydrotestosterone in rat prostate seems to be generally applicable. Shortly after the early work on the prostate gland was published, almost identical findings were reported in other male accessory sexual glands, especially rat epididymis (BLAQUIER, 1971; RITZÉN et al., 1971). The specific, high affinity binding of 5α-dihydrotestosterone in epididymal nuclei was reported by TINDALL et al. (1972) and the properties of the cytoplasmic receptor were extensively investigated by TINDALL et al. (1975). Reports by PODESTA et al. (1975) and ELKINGTON et al. (1975) suggest that the receptor mechanism and 5α-reductase required persistent androgenic stimulation for their maintenance at high operational activities. Specific means for distinguishing between the soluble androgen receptor and ABP, derived from the testis, have been developed by DANZO et al. (1975) and TINDALL et al. (1975). In a broad-based survey, MAINWARING and MANGAN (1973) established that androgen receptors, specific for 5α-dihydrotestosterone, were ubiquitous in male accessory sexual glands. Perhaps the outstanding feature of the study was the identification of such a receptor mechanism in adult rabbit prostate, for like the bull, this gland has a negligible ability to form 5α-dihydrotestosterone in the adult. Since 5α-reductase activity is readily demonstrated in the prostate of immature rabbits, then it seems likely that the androgen receptor mechanism is necessary for the period of prostate growth up to and including the attainment of sexual maturity (GLOYNA and WILSON, 1969; WILSON and WALKER, 1971). The adult rabbit prostate thus provides an example of an inactive or nonfunctional receptor mechanism; other examples are known (see section I.4d). The failure of 5α-androstane-3α, 17β- and -3β, 17β-diols to bind significantly to androgen receptors in many tissues of the rat was reported by KREIG et al. (1975); the 3α, 17β-diol had a limited capability to bind to some receptor proteins, but this was explainable by its conversion to 5α-dihydrotestosterone.

Androgen binding mechanisms in the human prostate gland provide potential avenues to improvements in the chemotherapy of prostatic carcinoma. To date, studies on the binding of androgens in human prostatic carcinoma are few (for example: MOBBS et al., 1974), but far more work has been done on nodules of human benign prostatic hyperplasia, beginning with the report by HANSSON et al. (1971). The most detailed study was conducted by MAINWARING and MILROY (1973) and this work has recently been corroborated by ROSEN et al. (1975). In all respects, the androgen-binding mechanism was seemingly identical to that described in rat prostate. The study by MAINWARING and MILROY (1973) remains distinctive on two accounts; first, the results of the binding of androgens *in vitro* was confirmed by experimentation *in vivo* where [^{3}H] testosterone (50 μCi) was administered as a single bolus 30 min before enucleation of benign prostatic adenomas and second, nuclear binding was studied in four samples of normal (nonadenomatous) prostate gland. However, the positive identification of specific, high affinity androgen receptors in surgical specimens of benign prostatic hyperplasia has not been reported by all investigators (STEINS et al., 1974; MOBBS et al., 1975). To a large extent, these apparent failures may be attributed to the intrinsic difficulties of working with hyperplastic nodules, for apart from their extreme resistance to homogenisation, they are heavily infiltrated with the plasma protein, SBG (STEINS et al., 1974; HANSSON et al., 1975). In sharp contrast to hopeful new developments in human breast cancer (FOLCA et al., 1961; MCGUIRE, 1973; SAVOLO et al., 1974; DE SOMBRE et al., 1974), quantitative determinations of androgen receptors seem to be of limited use in the diagnosis and clinical management of human prostatic hyperplasia and carcinoma. The human prostate gland seems to contain reproducible amounts of receptor protein and, as so lucidly described by GIORGI et al. (1973) and GIORGI (1976), the hyperplastic prostate gland of dog and man has a penchant for actively concentrating testosterone and its metabolites (see Chapter III.1a).

Of all the male accessory sexual glands, the dog prostate is the real anomaly in terms of androgen receptor mechanisms. Metabolic studies by HARPER et al. (1971) established that this gland can form appreciable quantities of testosterone metabolites, including 5α-androstane-3α, 17α-diol and this metabolite has the specific ability to enhance both DNA- and RNA-polymerase activities, *in vitro* (HARPER et al., 1970; DAVIES et al., 1972). While certain aspects of the stimulation of enzymes in cell-free systems by androgen metabolites remain far from understood, largely because androgenic responses can be evoked only in intact cells, EVANS and PIERREPOINT (1975) have now identified a cytoplasmic androgen receptor seemingly specific for 5α-androstane-3α, 17α-diol. This study provides an interesting comparison with rat prostate; the latter gland forms appreciable quantities of 5α-androstane-3β, 17α-diol, yet this stereoisomer is preferentially concentrated in microsomes rather than nuclei

Table 10. Binding of androgens in target cells other than rat prostate gland.

Organ	Optimum steroid for receptors	Reference	Nuclear binding	Reference	Metabolic capability of the organ	Reference
Rat epididymis	5α-Dihydrotestosterone	Blaquier (1971)	Yes	Tindall et al. (1972)	Abundant 5α-reductase	Wilson and Gloyna (1970)
Rabbit prostate	5α-Dihydrotestosterone	Mainwaring and Mangan (1973)	N.D. [a]	—	No 5α-reductase in adult	Wilson and Gloyna (1970)
Human prostate	5α-Dihydrotestosterone	Mainwaring and Milroy (1973)	Yes	Mainwaring and Milroy (1973)	Abundant 5α-reductase	Chamberlain et al. (1966)
Dog prostate	5α-Androstane-3α, 17α-diol	Evans and Pierrepoint (1975)	N.D.	—	Appropriate enzymes present	Harper et al. (1971)
Bone marrow	Testosterone	Minguell and Valladares (1974)	Yes	Valladares and Minguell (1975)	None required	—
Testis	{5α-Dihydrotestosterone Testosterone	Mainwaring and Mangan (1973) Mulder et al. (1975)	Yes	Smith et al. (1975)	No 5α-reductase in adult	Inano et al. (1967)
Skeletal muscle	Testosterone [b]	Michel and Baulieu (1974)	N.D.	—	5α-Reductase absent [c]	Mainwaring and Mangan (1973)

Skin fibroblasts	5α-Dihydrotestosterone	Keenan et al. (1975)	Yes	Keenan et al. (1975)	5α-Reductase present	Wilson (1975)
Mouse kidney	Testosterone [d]	Bullock and Bardin (1975)	Yes	Bullock and Bardin (1975)	None required	—
Anterior pituitary	{5α-Dihydrotestosterone Testosterone	Thieulant et al. (1975)	N.D.	—	5α-Reductase present (limited)	Thieulant et al. (1973)
Hypothalamus	5α-Dihydrotestosterone	Kato (1975)	Yes	Kato (1975)	5α-Reductase present	Rommerts and van der Molen (1971)
Cock comb	5α-Dihydrotestosterone	Dube et al. (1975)	N.D.	—	5α-Reductase present	Dube et al. (1975)
Mammary gland	5α-Dihydrotestosterone	Poortman et al. (1975)	N.D.	—	5α-Reductase present	Miller et al. (1974)
Uterus	Testosterone	Giannopoulos (1973)	Yes	Giannopoulos (1973)	None required	—
Vagina	Δ^5-Androstene-3 17β-diol	Shao et al. (1975)	Yes	Shao et al. (1975)	N.D.	—
Wolffian duct	Testosterone	Wilson (1973)	N.D.	—	5α-Reductase absent	Wilson and Lasnitzki (1971)
Foetal liver	Aetiocholanolone	Lane et al. (1975)	N.D.	—	5β-Reductase present [e]	Parsons (1970)

[a] *N.D.* = not determined; [b] receptors not always detectable by others; [c] exceptions include certain muscles in rabbit and guinea pig; [d] applies ***in vivo*** only; [e] plus other steroid metabolising enzymes.

(ROBEL et al., 1974). The canine prostate also forms 5α-dihydrotestosterone as a necessary precursor for the putative active metabolite, 5α-androstane-3α, 17α-diol; 5α-dihydrotestosterone is retained with a high affinity in the microsomes (KOWARSKI et al., 1969). For several reasons, therefore, the dog prostate is totally dissimilar to rat prostate in terms of androgen receptor mechanisms; the most outstanding difference being the formation and intracellular binding of a distinctive steroid, 5α-androstane-3α, 17α-diol. The studies by Pierrepoint and his collaborators raise serious objections to the tacit assumption that the ageing dog prostate provides a close experimental model for human prostatic carcinomas; the unequivocal demonstration of the nuclear binding of 5α-androstane-3α, 17α-diol is awaited with extreme interest.

The binding of androgens in bone marrow has two interesting features; first, the binding mechanism is selective for nonmetabolised testosterone and second, a cytoplasmic androgen receptor could not be identified, despite the fact that the nuclear binding of [^{3}H] testosterone was reproducibly demonstrated (MINGUELL and VALLADARES, 1974). However, these authors concluded that an androgen-binding protein was extractable from purified marrow nuclei in 0.5 M KCl. This observation was extended by VALLADARES and MINGUELL (1975) who rigorously examined the properties of the nuclear binding component; sufficient steroid specificity was present in the nuclear extracts to justify the identification of an androgen receptor. The single class of nuclear binding sites were saturated at 1×10^{-8} M testosterone and their high affinity was confirmed by the apparent K_d of 5.9×10^{-9} M. The 5α- and 5β- stereoisomers of dihydrotestosterone competed sluggishly for these nuclear sites and no displacement was found with oestradiol-17β. This system is almost unique among androgen target cells in displaying specific nuclear binding of high affinity without any apparent involvement of a cytoplasmic receptor protein; one more example is cited below. A similar absence of cytoplasmic receptors exists in the nuclear binding of oestrogens in foetal chick liver (MESTER and BAULIEU, 1972) and of aldosterone in embryonic kidney (PASQUALINI et al., 1972). The results of nuclear binding in bone marrow contrast sharply with similar studies on nuclear extracts of rat prostate; in the latter, binding was found *in vitro* but it was nonspecific and of low affinity (MAINWARING, 1969 b).

Since it has long been recognised that androgens have a direct role in spermatogenesis (for review, see STEINBERGER, 1971), it is surprising that androgen receptor mechanisms have only recently been studied in depth. MAINWARING and MANGAN (1973) described a high affinity, androgen-binding protein in rat testis but the identification of a cytoplasmic androgen receptor must be credited to HANSSON et al. (1974). The testicular receptor was similar to that in rat prostate and epididymis but distinct from the non-receptor protein, ABP. This testicular protein merited description as a receptor and was distinguishable from ABP in terms of electrophoretic mobility, temperature stability, and rates of dissociation of optimum ligands. The receptor in testis is distinctive in that it binds testosterone and 5α-dihydrotestosterone with a similar affinity (HANSSON et al., 1974); this ambivalence in binding specificity distinguishes it from the androgen receptor in rat prostate. Additional work by MULDER et al. (1975) and SMITH et al. (1975) established that both testosterone and 5α-dihydrotestosterone could be retained in testicular nuclei after the injection of [^{3}H] testosterone, in vivo; both of these reports noted a sensitivity of the binding process to cyproterone acetate and the only difference was the preferential concentration of testosterone in the nucleus reported by MULDER et al. (1975). As discussed in Chapter II.3a, the rat testis provides a further example of dramatic changes in 5α-reductase activity during development; as a consequence, the binding of 5α-dihydrotestosterone can only be significant during the early phases of adult growth up to the attainment of sexual maturity. Testosterone may be required only for the maintenance of testicular growth. It is not yet clear whether these two androgens have distinct functions in the prostate, but some indication could be obtained by use of inhibitors of 5α-reductase at various ages of the developing rat. Future work must be directed towards two principal objectives; first, do two receptors exist in testis or only one with a broad steroid specificity and second, do the complexes with testosterone and 5α-dihydrotestosterone modify the structure of the receptor protein such that it is retained at different sites within nuclear chromatin?

As so lucidly emphasized by KOCHAKIN (1975), the molecular basis for the anabolic function of androgens in skeletal muscle remains something of a puzzle. The binding of androgens has been studied in skeletal muscle and also bulbocavernosus muscle, a particularly androgen-sensitive muscle, derived from the embryonic perineal complex; prompted by the work of HAYES (1965), it is more correct to use this terminology than the older name, levator ani muscle. With the possible exception of the bulbocavernosus muscle of guinea pig (MAINWARING and MANGAN, 1973), most investigators agree that skeletal muscle has a minimal capability to form 5α-dihydrotestosterone (see Chaper II.2). Unless putative androgen receptors in muscle can function at a much lower threshold of androgenic stimulation than male accessory sexual glands, the high-affinity binding of this active metabolite is unlikely to be implicated in the anabolic function of androgens. The failure to demonstrate significant binding of testosterone in skeletal muscle, including bulbocavernosus muscle, has been widely documented (GLOYNA and WILSON, 1969; WILSON and GLOYNA, 1970; MAINWARING and MANGAN, 1973; KREIG et al., 1974). Although a relatively higher uptake of [^{3}H] labelled androgens was noted by KREIG et al. (1974) compared with other muscles, specific high-affinity complexes containing bound androgen ligands could not be identified by agar gel electrophoresis. Other evidence is diametrically opposed to these negative findings. POWER and FLORINI (1975) reported that testosterone rather than 5α-dihydrotestosterone promoted DNA synthesis in cultures of muscle cells in vitro, and testosterone-specific receptors have been identified in muscle extracts by JUNG and BAULIEU (1972) and MICHEL and BAULIEU (1974). From a close examination of all papers on the binding of androgens in muscle, the only distinctive feature of the study by MICHEL and BAULIEU (1974) was the use of muscle preparations containing a high concentration of protein; this may have afforded some measure of protection if the muscle receptors were particularly labile proteins. In the present author's view, there is no real explanation of these conflicting findings, but my personal prejudice is that the androgen binding mechanism in muscle, if it indeed exists, is quite different in character to that in rat prostate. As supporting evidence, the following facts are presented. First, it is possible that the binding moieties may be extremely labile or present in quantities below the detection limits of current methods; this seems a little unlikely, since the negative reports stem from several independent laboratories. Second, a clear distinction may be drawn between the structural determinants present in androgenic as compared with anabolic steroids (see Fig. 2.) and a differentiation has been made between the androgenic and anabolic binding sites in muscle (STEINITZ et al., 1971). In this study, dexamethasone could negate only the anabolic binding sites; this is not surprising, since this glucocorticoid fails to compete for the androgen-binding sites in all other androgen target cells. Third, it is arguable that the anabolic sites in muscle have a fundamentally different structure to other binding sites for androgen-related steroids and thus may not be readily solubilised. To illustrate the possibility of uniqueness of structure, growth hormone has a synergistic effect to testosterone on muscle (SCOW and HAGEN, 1965) and binding assays should be repeated on hypophysectomised animals with the concomitant injection of pituitary hormones with [^{3}H]-labelled anabolic steroids. To clarify these contentious issues, the best course is to monitor the binding of synthetic steroids, such as norbolethone, which have a high anabolic : androgenic ratio. This approach should at least identify the putative anabolic sites, but if the experimental findings were negative, then the sites could be different in fundamental structure or beyond the limits of detection by conventional means. Above all, the appearance of [^{3}H]-labelled norbolethone in muscle nuclei would indicate the presence of some form of selective binding mechanism.

Receptors for both androgens and oestrogens have been identified in rat skin (EPPENBERGER and HSIA, 1972). These high-affinity binding proteins differ in molecular size and also in their relative concentrations during the hair cycle; oestradiol binding is maximal at the resting (telogen) phase but testosterone binding reaches its peak during the transition (catagen) phase. A 5α-dihydrotestosterone-binding system identical to that in male accessory sexual glands has been identified by ADACHI and KANO (1972) in the androgen-sensitive sebaceous or costovertebral glands of the male Syrian hamster. The most comprehensive description of an androgen receptor system in skin and its associated structures is by KEENAN et al. (1975) on androgen-sensitive fibroblasts derived from human skin. Even after protracted culture, these fibroblasts maintain their ability to bind 5α-dihydrotestosterone with a high affinity (apparent K_d 0.2 to 1.6×10^{-9} M). These receptor sites were conspicuously enriched in cells derived from the perineal skin of the newborn, but were also readily identified in skin from other areas, even from adult donors. Receptor proteins were present in both the cytoplasmic and nuclear compartments and as far as can be judged on present evidence, the binding mechanism in skin fibroblasts is identical to that described in rat prostate. Saturation analysis indicates a maximum of 18,000 specific sites for 5α-dihydrotestosterone per fibroblast.

The binding of androgens in brain is a very complex process indeed. Using the technique of dry-mount autoradiography, SAR and STUMPF (1973 a) located the radioactivity in brain after the administration of [^{3}H] testosterone to immature or castrated rats, *in vivo*. Retention of tritium was found in many areas of the brain associated with the control of gonadotrophin secretion, including the nucleus (or, simply, n.) arcuatus, n. ventromedialis hypothalami, n. preopticus, n. septi lateralis, hippocampus, and amygdala. By the same procedure, SAR and STUMPF (1973 b) looked for the selective retention of androgens in rat pituitary. The nuclei of the gonadotrophin-secreting basophils of the pars distalis of the anterior pituitary actively retained [^{3}H]-labelled androgens, but this was not evident in the intermediate or posterior lobes. This morphologic evidence is exceeding clear-cut, but purely biochemical approaches have been less successful. From studies both *in vivo* and *in vitro*, considerable evidence has been accumulated that 5α-dihydrotestosterone, 5α-androstane-3α, 17β-diol and 5α-androstanedione may be formed in the anterior and posterior hypothalamus (SHOLITON et al., 1972; WHALEN and RESEK, 1972). However, a higher accumulation of [^{3}H]-labelled androgens in brain over their concentration in plasma is only strikingly demonstrable when other androgen-binding sites are severely reduced, as in functionally hepatectomised or totally eviscerated animals (SHOLITON et al., 1972). It is perhaps for this reason that others have failed to demonstrate a significant uptake of [^{3}H] testosterone into the brain (BARNEA et al., 1972; TUOHIMAA and NIEMI, 1972). The binding of androgens in rat anterior pituitary is now more clearly understood. The enzyme, 5α-reductase, is present in only limited amounts in the anterior pituitary (THIEULANT et al., 1973) and two testosterone-binding components were identified by JOUAN et al. (1973) by gel-exclusion chromatography on Sephadex G-200. This study indicated a very high background of nonspecific androgen binding in pituitary extracts, and this has hampered progress toward the characterisation of the authentic androgen receptor components. A cytoplasmic binding protein, said to be specific for testosterone, was first detected by SAMPAREZ et al. (1974) and then a high affinity, 5α-dihydrotestosterone-binding protein was discovered as well (THIEULANT et al. (1975), apparent K_d 7.8 × 10^{-10} M. Detailed studies on the steroid specificity of the binding proteins indicate that only testosterone and 5α-dihydrotestosterone are bound to any appreciable extent and they may thus be termed androgen receptors. It still remains uncertain, however, whether there is a simple protein binding both testosterone and its active metabolite, or two proteins with an elective preference for one specific androgen. More work is needed to clarify this matter and nuclear binding studies should also be conducted. Binding studies on hypothalamus are more advanced and certainly 5α-reductase is present in this area of the brain (ROMMAERTS and VAN DER MOLEN, 1971; NOMA et al., 1975). The enzyme is distributed throughout all subcellular fractions and is seemingly unaffected by castration or androgenic stimulation; in these respects, the hypothalamic 5α-reductase differs from its counterpart in rat prostate. A detailed study of the binding of 5α-dihydrotestosterone in hypothalamus has been performed by KATO (1975) and this creditable work has helped to clarify the role of androgen binding in the feedback mechanisms in the brain necessary for controlling the release of gonadotrophins. A thermolabile, 8S androgen receptor was present in soluble extracts of hypothalamus with a high affinity for 5α-dihydrotestosterone, apparent K_d 0.7 to 2.4 × 10^{-9} M. This specific binding was reduced only by oestradiol-17β, testosterone, and cyproterone acetate; other steroids were ineffective competitors. The androgen receptor was located in the median eminence and of outstanding importance, it was present in all experimental rodents and showed a developmental increase in concentration, reaching a maximum in the rat 28 days after birth. This is precisely the time when sexual maturity is attained. Since specific nuclear binding of androgens was also established (KATO, 1975), the overall binding scheme in hypothalamus is similar, if not identical, to that in rat prostate. In summary, the binding of androgens in the brain remains a complex but fascinating problem. Both tesosterone and 5α-dihydrotestosterone are bound with a high affinity in the anterior pituitary and hypothalamus; what is clearly needed now is unequivocal evidence favouring either one or two androgen receptor systems in the brain, specific for either testosterone or 5α-dihydrotestosterone. The important contribution by KATO (1975) indicates the involvement of androgen-binding mechanisms in the differentiation and maintenance of male sexual characteristics.

Since the comb and wattles of the capon played such a distinctive part in the historical background of experimental research on androgenic hormones, it is a little surprising that androgen-binding mechanisms have been identified in these androgen target tissues only within the last year. Androgen receptors are present in both comb and wattles (DUBÉ et al., 1975) and these are possibly specific for 5α-dihydrotestosterone. The physicochemical properties of these binding proteins have not been described. Similar considerations apply to the androgen receptors in human mammary carcinomata (POORTMAN et al., 1975) which undoubtedly contain high levels of 5α-reductase activity (MILLER et al., 1974).

In two organs of the reproductive tract of female rats, the vagina and uterus, the high-affinity binding of testosterone and its metabolites has been unequivocally identified (GIANNOPOULOS, 1973; SHAO et al., 1975). In uterus, the binding mechanism is a total reflection of that in male accessory sexual glands, including the temperature dependence of the nuclear binding process and the absolute necessity for a high-affinity, cytoplasmic receptor protein. The interesting feature of this system is that testosterone is the optimum ligand and the binding mechanism is insensitive to cyproterone acetate (GIANNOPOULOS, 1973). In rat vagina, the selectively bound steroid is a metabolite of testosterone, Δ^5-androstenediol, and the distinctive aspect of this binding process is the direct retention of the active metabolite in nuclei without the implication of a cytoplasmic receptor protein (SHAO et al., 1975). This indicates a truly novel type of binding mechanisms, and as mentioned earlier for bone marrow, the physical properties of these unique nuclear proteins should be determined as a matter of priority. Unfortunately, there is no information on the competition by cyproterone acetate for these nuclear sites in vagina. A soluble receptor, specific for 5α-dihydrotestosterone, has been identified in calf uterus by MCCANN et al. (1970). This is an interesting interspecies difference, as rat uterus, which preferentially binds nonmetabolised testosterone, does not contain 5α-reductase activity (GIANNOPOULOS, 1973). These androgen receptors may be germane to the differential effects of androgens and oestrogens on uterine function (GONZALEZ-DIDDI et al., 1972).

Androgens are specifically bound with a high affinity in the kidneys of both sexes of genetically inbred strains of mice. The only area of debate with this androgen-binding mechanism is its specificity with respect to 5α-dihydrotestosterone and indeed, the authentic presence of an active 5α-reductase system. When assayed under conditions *in vitro*, 5α-reductase activity was detected in kidney minces by VERHOEVEN and DE MOOR (1971) and MAINWARING and MANGAN (1973) but not by MOWZOWICZ and BARDIN (1974). Again under conditions *in vitro*, the cytoplasmic binding of both testosterone and 5α-dihydrotestosterone was reported (GEHRING et al., 1971; MAINWARING and MANGAN, 1973). On the other hand, when [^{3}H] testosterone was administered to mice *in vivo*, either alone or in the additional presence of nonradioactive steroid competitors, tissue- and steroid-specific binding was detected in kidney nuclei. Furthermore, the nuclear radioactivity was recovered primarily as nonmetabolised [^{3}H] testosterone (BULLOCK et al., 1971; BULLOCK and BARDIN, 1975 a). In their recent study, BULLOCK and BARDIN (1975 a) scrupulously addressed the specificity of the nuclear binding of [^{3}H]-labelled steroids in the kidney of Balb/C mice. Only [^{3}H] testosterone was recovered in kidney nuclei when the [^{3}H]-steroid administered was testosterone or androstenedione; in contrast, predominantly [^{3}H] 5α-dihydrotestosterone was retained in the nuclei when the [^{3}H] tracer steroid was 5α-dihydrotestosterone or 5α-androstane-3α (or 3β), 17β-diol. With both [^{3}H] steroid precursors, competition for the nuclear binding sites was observed with nonradioactive 5α-dihydrotestosterone, testosterone, or cyproterone acetate. Under identical injection protocols, [^{3}H] progesterone was not retained in kidney nuclei; [^{3}H] cortisol was bound, however, but the binding of this potent glucocorticoid was not curtailed by the injection of nonradioactive androgens. In summary, there have been doubts concerning the activity of 5α-reductase in mouse kidney and these conflicting findings may be a reflection of methodologic differences from one laboratory to another, particularly in the provision of the added coenzyme, NADPH. However, there is no doubt that under conditions *in vivo,* there are specific nuclear binding mechanisms in mouse kidney for both androgens and glucocorticoids. The outstanding property of the androgen binding mechanism is its high affinity for both testosterone and 5α-dihydrotestosterone, but testosterone is the favoured androgen under physiological conditions (BULLOCK and BARDIN, 1975 a).

Despite this wealth of detail, there are large areas of weakness in our knowledge of the binding of androgens. Testosterone or its active metabolites are needed for the differentiation of chick oviduct (YU and MARQUARDT, 1973), the maturation of Graafian follicles in rat ovary (LOUVET et al., 1975), and the uptake and utilisation of glucose in many androgen target cells (for example, HÄRKÖNEN et al., 1975). In these instances, receptor mechanisms have either not been identified or their involvement remains to be vigorously verified. Taking a more general viewpoint, the paucity of information on androgen-binding mechanisms in foetal tissues is striking, a notable exception being the work of WILSON (1973) on the binding of testosterone in the Wolffian duct. In a similar vein, work on neonatal tissues is extremely fragmentary, especially in seeking plausible explanations for such phenomena as androgenic programming, enzyme imprinting, and neonatal androgenisation. It is to be hoped that these important developmental aspects will feature more prominently in future research on androgen-binding mechanisms. As emphasized by contemporary studies on the brain, investigators must adopt a more critical appraisal of androgen binding and be less satisfied with binding for its own sake; the specificity and identity of many receptor mechanisms must be put on a sounder footing if the subtle biology of many important systems is to be fully understood.

b) Physical Properties of Cytoplasmic and Nuclear Receptors

From the summary presented in Table 10, androgen receptors have been identified and characterized in a remarkable diversity of androgen target cells. Bearing in mind the marked contrasts in the androgenic responses and the nature of the optimum ligands formed *in vivo,* it is perhaps surprising that the cytoplasmic receptors are so similar in their physicochemical characteristics. Sedimentation coefficients of approximately 8S have been reported for cytoplasmic receptors from such diverse origins as rat epididymis (TINDALL et al., 1975), hypothalamus (KATO, 1975), uterus (GIANNOPOULOS, 1973), mouse kidney (BULLOCK et al., 1975), human prostate (ROSEN et al., 1975), and Shionogi 115 tumours (MAINWARING and MANGAN, 1973). Whether this 8S form represents the true configuration of the receptors in intact cells remains in question, but where they have been identified, they are dissociated in 0.5 M KCl to a form sedimenting at 4–5S. Under conditions of low ionic strength, other androgen receptors have a 4S form (KEENAN et al., 1975; EVANS and PIERREPOINT, 1975). Whether this difference is real or simply a reflection of preparative methods is open to conjecture. Generally speaking, receptors are acidic proteins. A pI of 5.8 was noted for cytoplasmic receptors in all male accessory sexual glands and testis (MAINWARING and IRVING, 1973; MAINWARING and MANGAN, 1973; TINDALL et al., 1975); even lower values were noted for receptors in mouse kidney (pI 4.8; BULLOCK et al., 1975) and bone marrow (pI 4.9; VALLADARES and MINGUELL, 1975). There is a consensus of opinion that all receptors are thermolabile, being rapidly inactivated at temperatures above 30° C and equally sensitive to blockade of essential -SH groups. Receptor complexes have high molecular weights in the range 120,000 to 270,000, depending on the method and conditions of analysis and frictional ratios in the range 1.60–1.74.

The nuclear forms of androgen receptor complexes have not been studied in sufficient detail to permit any generalisations except that they are proteinaceous and invariably released by extraction into 0.5 M KCl. Apart from the rat prostate gland, the activation or transformation of cytoplasmic receptor complexes has received little attention and the development of reconstituted systems for nuclear transfer experiments is not generally advanced. The nuclear androgen receptors in rat vagina (SHAO et al., 1975) and bone marrow (VALLADARES and MINGUELL, 1975) deserve closer scrutiny of their physicochemical properties because they are able to bind their optimum ligands without any apparent involvement of cytoplasmic receptor proteins. This property makes them unique among androgen target cells.

Despite their selective propensity for testosterone or active metabolites (or indeed, both androgens, as in mouse kidney; BULLOCK and BARDIN, 1975 a), the high affinity binding sites must have much in common at the molecular level. Without exception, the receptors in androgen target cells have no affinity for glucocorticoids or mineralocorticoids; progestational steroids are also generally bound with a relatively low affinity. Cyproterone acetate also has an almost ubiquitous ability to negate the high-affinity binding of androgens; notable exceptions include the testosterone-specific receptor in rat uterus (GIANNOPOULOS, 1973) and the as yet poorly defined binding mechanism responsible for neonatal programming (BRONSON et al., 1972; DIXIT and NIEMI, 1973). Of all androgenic responses, only the induction of kidney β-glucuronidase (OHNO and LYON, 1970) and pentose shunt enzymes in rat prostate (MANGAN et al., 1973) are insensitive to antiandrogens.

c) Binding of Hormones Other than Androgens

This topic is somewhat divorced from the main theme of this monograph, but to illustrate the complexity of hormone-induced phenomena as reflected in high-affinity binding mechanisms, certain reference must be made. Mouse kidney is able to concentrate [^{3}H] cortisol in both the nuclear and cytoplasmic compartments (BULLOCK and BARDIN, 1975 a) and a specific oestrogen receptor mechanism is also present

(BULLOCK and BARDIN, 1975 b). The oestrogen receptor is of interest because it persists even in androgen-insensitive Tfm mice, indicating that oestrogen and androgen receptors are under independent genetic control. Oestrogen receptors are also detectable in the cytoplasmic and nuclear fractions of rat testis (BRINKMANN et al., 1972; MULDER et al., 1973) and they are also a prominent feature of the hypothalamus and pituitary of both male and female animals (KATO, 1973; PLAPINGER and MCEWEN, 1973; KATO et al., 1974; KORACH and MULDOON, 1974). These oestrogen receptors are almost certainly involved in feedback control mechanisms, and in the brain help to illustrate the unique binding of androgens and oestrogens in the same target cells. These examples provide some idea of the extent and complexity of receptor mechanisms in mammalian cells.

d) Specificity of Androgenic Responses

In Chapter I.4d, the acute tissue specificity of androgenic responses was raised; examples included the restricted induction of alcohol dehydrogenase in mouse kidney (OHNO et al., 1970) and the unique regulation of basic proteins in rat seminal vesicle (TÓTH and ZAKÁR, 1971) by androgens. The critical questions, then, are how these selected processes are initiated and to what extent cytoplasmic receptors and nuclear acceptor proteins are involved in the maintenance of tissue specificity. To provide a platform for the ensuing discussion, specificity is expressed in two categories, termed here overt and discrete specificity. Overt specificity describes the sensitivity or refractoriness of cells to steroid hormones. Discrete specificity describes the subtle differences in the responses of many tissues to the same hormone. Because they represent the first real determinant in the expression of cellular responses, it is reasonable to ask if the cytoplasmic receptor proteins are themselves tissue-specific. Until they may be purified to homogeneity, the only valid indication of tissue specificity in receptor structure remains the differences in their binding affinity for steroid ligands. On this criterion, absolute specificity is currently limited to the glucocorticoid receptors in different areas of the brain (DE KLOET et al., 1975), the progesterone receptors in uteri from different species (KONTULA et al., 1975) and perhaps the androgen receptor in rat uterus (GIANNOPOULOS, 1973).

To appraise hormonal specificity critically, MAINWARING et al. (1976) prepared receptor complexes and acceptor components from many hormone-sensitive tissues, including rat liver, rat uterus, and a wide spectrum of male accessory sexual glands. These preparations enabled "cross-over" experiments to be undertaken, using the procedure of affinity chromatography (see section III.1e) and the immobilisation of the acceptor components to Sepharose 2B. Considering overt specificity first, this can largely be explained by the binding specificity of cytoplasmic receptor proteins, assisted to a certain extent by the selectivity of acceptor proteins for retaining receptor complexes. The best example of this is seen in the interchange of preparations from androgen- and glucocorticoid-sensitive cells. The androgen receptor of rat prostate does not bind dexamethasone and the glucocorticoid receptor of rat liver does not bind 5α-dihydrotestosterone. When receptor complexes were applied to acceptor-containing matrices, the acceptor sites recognised only the homologous receptor complex or that of identical origin; the specific acceptor sites in prostate preparations could only be occupied by androgen receptor complex and vice versa. To investigate discrete specificity, similar interactions were conducted on receptor and acceptor preparations derived from many male accessory sexual glands. Discrete specificity cannot be explained simply by receptor-acceptor interactions alone, at least not by present methods (see YAMAMOTO and ALBERTS, 1975, for review). While a given androgen receptor complex showed some preference in binding more extensively to its homologous acceptor matrix, very extensive association with heterologous acceptor sites was observed (MAINWARING et al., 1976 b). This limitation was exposed by the interchange of subcellular preparations derived from prostate, epididymis, and preputial gland.

This work, along with the conclusions of YAMAMOTO and ALBERTS (1975), raises the fundamental problem which will probably transcend other aspects of binding mechanisms in the foreseeable future; stated quite simply, how is tissue specificity maintained at the molecular level? Although we now have a clearer idea of the structure of chromatin, present methods for analysing acceptor-receptor interactions are not sufficiently precise to distinguish the truly tissue-specific sites in the genetic apparatus. All that can be said at the present time is that during the process of cellular differentiation, specificity is imparted to androgen target cells at the level of the cytoplasmic receptor protein and also in the structure of the acceptor sites in chromatin.

KING and MAINWARING (1974) discussed the theoretical possibilities of receptor-acceptor interactions from the standpoint of tissue specificity; such interactions fall into two classes, encompassing either direct or indirect effects on chromatin DNA (Fig. 8). The indirect effects include two possibilities. (a) The acceptor is a tissue-specific, nonhistone protein which undergoes a conformational change after interaction with the receptor complex. This modifies the regulatory function of the acceptor and the characteristic response of the target cells is initiated. (b) By interaction with the acceptor protein, the receptor is located at the correct, tissue-specific region of chromatin and a secondary reaction with a histone then occurs, allowing the tissue-specific region of chromatin to be exposed or activated. There are also two possible mechanisms involving direct effects. (c) Specific interaction between the acceptor and receptor complex involves a secondary reaction with DNA. (d) This implies that chromosomal proteins modify the structure of the genome in that tissue-specific sites on DNA are exposed in target cells but blocked in nontarget cells. This scheme does not involve any tissue-specific reaction between receptor complex and acceptor proteins. Among androgen target cells, there is no really convincing evidence favouring any of these four alternatives; scheme (a) is probably best supported by currently available experimental findings.

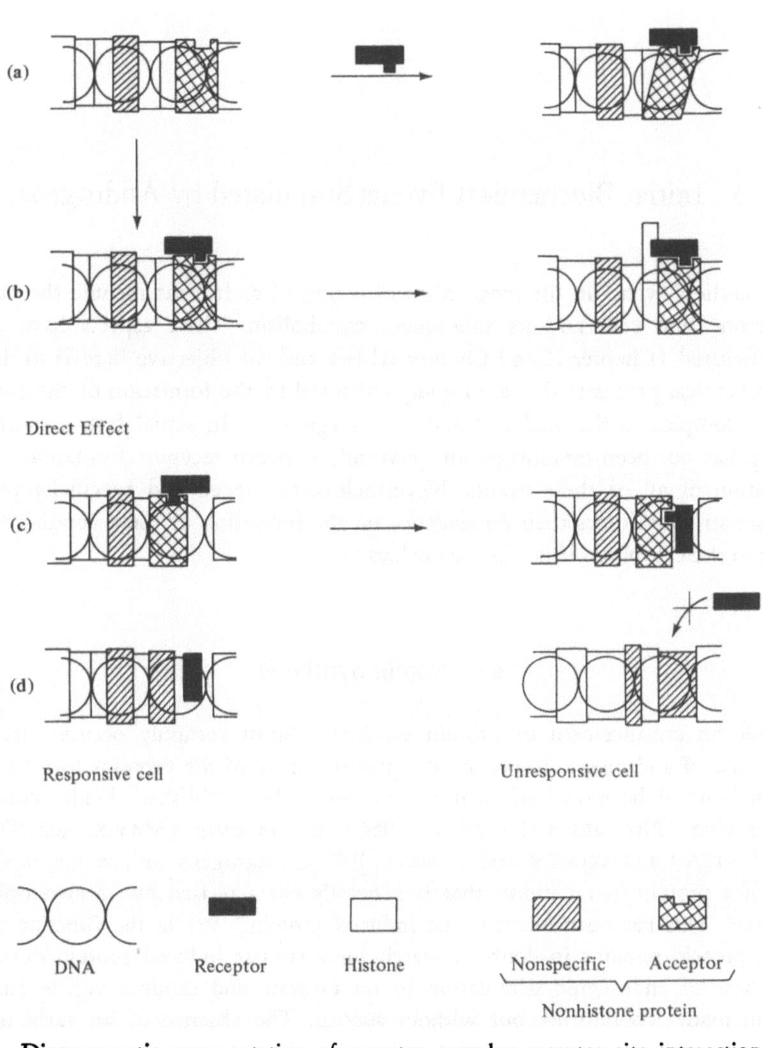

Fig. 8. Diagrammatic representation of receptor complex-acceptor site interactions. Figure is taken from KING and MAINWARING (1974) and each of the four alternatives is described fully in text

3. Initial Biochemical Events Stimulated by Androgens

The earliest events in the mechanism of action of androgens include the entry of testosterone into cells and its subsequent metabolism. These aspects have already been discussed (Chapter II and Chapter III.1a) and the objective here is to describe the biochemical processes that are rapidly enhanced by the formation of the androgen receptor complex in the nuclei of androgen target cells. In actual fact, a causal relationship has not been established unequivocally between receptor formation and the stimulation of all of these events. Nevertheless, the speed and specificity of these processes strongly infers their dependence on the formation of the receptor complex, at least in the cytoplasm if not in the nucleus.

a) Protein Synthesis

While an enhancement of protein synthesis almost certainly occurs during the rapid phase of androgenic responses, the precise nature of the proteins formed within the first hour of hormonal stimulation remains to be established. Under conditions both in vivo (NOTIDES and GORSKI, 1966) and *in vitro* (MAYOL and THEYER, 1970; KATZENELLENBOGEN and GORSKI, 1972), oestrogens induce the rapid synthesis of a protein in rat uterus that is generally characterised by its electrophoretic properties. This has been termed "the induced protein." While the function of this uterine protein remains in doubt, a search for a similar induced protein during the early phase of androgenic stimulation in rat prostate and seminal vesicle has been made in many laboratories, but without success. The absence of an early induced protein in male accessory glands has been confirmed recently in the study by PENNEQUIN et al. (1975). Thus, the proteins made during the early phase of androgenic responses are unknown.

b) Activation of Protein Phosphokinases

When the basic charges on nuclear histones are blocked by phosphorylation, their ability to bind to DNA is severely impaired. The chemical modification of histones changes the structure of chromatin and it is widely believed that this process is involved in the activation of the template properties of chromatin for RNA polymerase. AHMED (1971) described a protein phosphokinase in purified rat prostate nuclei that was capable of transferring the γ-phosphate of ATP into histones. This interesting study strongly suggests that this enzyme may be involved in the very early stages of the androgenic response. The prostate nuclear phosphokinase was depleted after castration, yet on the administration of testosterone to castrated rats, a significant enhancement of activity was detected within 30 min. While the steroid specificity of this induction was not explored, the nuclear phosphokinase in liver was not reduced after castration and was insensitive to androgenic stimulation. The critical feature of this study (AHMED, 1971) was that the addition of testosterone or 5α-dihydrotestosterone to purified nuclei, even at concentrations as high as 10^{-4} M, had no influence on the activity of nuclear phosphokinase. The androgenic response could only be elicited in the intact cell, making it highly likely that the enzyme was activated by cytoplasmic proteins and perhaps even the androgen receptor complex. Somewhat later, multiple protein phosphokinases were found to be an integral feature of nuclear chromatin (TAKEDA et al., 1971; KISH and KLEINSMITH, 1974) and WILSON and AHMED (1975) have now described discrete protein phosphokinases in both the nucleolar and extranucleolar fractions of rat prostate nuclei. The nucleolar enzyme is particularly interesting, because a stimulation of nucleolar RNA synthesis is one of the earliest and most extensive changes evoked by androgens (see section III.4d). It appears, then, that the phosphorylation of histones and other nuclear proteins is a very early event in the androgenic response in rat prostate. This mechanism could promote localised changes in the structure of chromatin, thus allowing the transcription of distinct areas of the genome. Two aspects of prostate nuclear phosphokinases warrant attention in the future. First, the steroid specificity of this induction should be clarified, particularly by the use of antiandrogens. Second, it is of great importance to establish whether cytoplasmic proteins, including the androgen receptor complex, can stimulate the chromatin-associated protein phosphokinases in reconstituted, cell-free systems.

c) Stimulation of the Initiation Step of Protein Synthesis

In recent years, much has been learned about the early steps of polypeptide synthesis in eukaryotic cells (for a review, see LUCAS-LENARD and LIPMANN, 1971). A complex between *L*-methionine and a specific species of transfer RNA, $tRNA_f^{Met}$, is responsible for the initiation of all polypeptide chains from their N-terminal end and the complex associates with ribosomes, provided that GTP and specific protein factors are present. LIANG and LIAO (1975 a, b) have made exciting progress in demonstrating how androgens may possibly regulate the concentration or activity of protein initiation factors in rat prostate. Their assay depends on the association of [^{35}S] methionyl-$tRNA_f$ with prostate ribosomes and the retention of the labelled complex on nitrocellulose membranes. After castration, the amount of cytoplasmic initiation factors in rat prostate is reduced, yet may be stimulated within 10 min of the administration of testosterone, *in vivo* (LIANG and LIAO, 1975 a, b). The comments made above on nuclear phosphokinases are equally applicable to the regulation of cytoplasmic initiation factors. It is essential to establish the tissue- and steroid-specificity of these effects. In addition, the possibility that the cytoplasmic androgen receptor complex is involved in this rapid response should also be pursued.

d) The Androgenic Regulation of RNA Synthesis: Receptor Effects *in vitro*

It has long been recognised that androgens stimulate the synthesis of RNA in androgen target cells (for reviews, see WILLIAMS-ASHMAN and REDDI, 1971; and VILLEE et al., 1975). Some of these androgen-mediated effects are evident within the first 30 min of androgenic stimulation. Rapid responses in RNA synthesis were noted by WICKS and KENNEY (1964) in rat seminal vesicle, where the incorporation of ^{32}P into total RNA was doubled within 20 min of the injection of testosterone into castrated rats. Work by LIAO et al. (1966) and LIAO and LIN (1967) indicated that ribosomal RNA was rapidly stimulated after androgenic stimulation and this was confirmed by MAINWARING et al. (1971) by studies on purified nucleoli of rat prostate.

Work on these early responses in RNA synthesis would be expedited if androgens could evoke their biological responses in target cells, *in vitro*. However, ALFHEIM and FJELL (1973) failed to demonstrate an increase in the synthesis of RNA in prostate minces *in vitro* in the presence of 10^{-6} M 5α-dihydrotestosterone. More recently, it has been established that rat epididymal tubules may be cultured successfully in chemically defined media and maintain their response to 5α-dihydrotestosterone in terms of an acceleration of RNA synthesis (BLAQUIER et al., 1975). This experimental system may be more amenable for studies on rapid biochemical responses. It had been claimed in the past that the incorporation of radioactive precursors into RNA in purified nuclei could be stimulated directly by testosterone and 5α-dihydrotestosterone (BASHIRELAHI et al., 1969; BASHIRELAHI and VILLEE, 1970). These observations could not be confirmed by other investigators (MAINWARING et al., 1971). It would appear that an intact androgen target cell is necessary for the enhancement of RNA synthesis by androgens and this implies the involvement of cytoplasmic proteins, including the androgen receptor complex.

This early activation of RNA synthesis could be explained by several processes; activation of the chromatin template, the selective sequestering of essential divalent cations, the stimulation of the supply of obligatory precursors such as the nucleoside 5′-triphosphates or activation of the enzyme, RNA polymerase. No information is currently available on some of these possibilities, but opinion is tending to favour strongly the activation of the chromatin template. This trend illustrates the extreme importance of the work on protein phosphokinases (AHMED, 1971; WILSON and AHMED, 1975) described earlier in section III.3b. As reviewed by JACOB (1973), eukaryotic RNA polymerases have been extensively characterised and it is possible that both the nucleolar (form A; amanitin-insensitive) and extranucleolar (form B; amanitin-sensitive) forms of RNA polymerase work more efficiently on chromatin template DNA after androgenic stimulation. MAINWARING et al. (1971) solubilised and fractionated rat prostate RNA polymerases but no studies are available on these enzyme profiles in the early phase of the androgenic response. The remaining discussion will be restricted to the putative activation of the chromatin template by androgens.

The activation of nucleolar chromatin by the androgen receptor complex must be elicited by indirect means, because the receptor is barely demonstrable in rat prostate nucleoli (MAINWARING, 1969 b). The activation of the nucleolar RNA polymerase could theoretically be achieved by conformational or allosteric changes but there is no experimental evidence for this. A more plausible explanation is that the receptor triggers the synthesis of a regulatory protein in the extranucleolar region of the nucleus, which in turn migrates into the nucleolus and activates the DNA template or enables form A RNA polymerase to bind to DNA more effectively. The existence of such regulatory proteins in eukaryotic cells has been widely described (WILLEMS et al., 1969; MURAMATSU et al., 1970; WANKA and SCHRAUWEN, 1971).

The activation of total (extranucleolar) chromatin by androgens has been reported by LIAO and LIN (1967) and MANGAN et al. (1968). However, in both of these studies, the template activity of chromatin was measured by the synthesis of RNA in the presence of highly purified RNA polymerase from *E. coli* or *Micrococcus luteus*. This tends to cloud the importance of these early studies, because there are now serious doubts as to the authenticity of template assays conducted with prokaryotic RNA polymerases. This point was first raised by KESHGEGIAN and FURTH (1972) who established that *E. coli* polymerase bound somewhat randomly to mammalian chromatin and transcribed RNA at most of these nonspecific promotor sites; by contrast, mammalian polymerase bound to a restricted number of selective sites and actively transcribed them all. The particular fidelity of the transcription of chromatin by mammalian rather than bacterial RNA polymerase has since been widely substantiated (REEDER, 1973; MARYANKA and GOULD, 1973; STEGGLES et al., 1974). MAINWARING et al. (1971) measured the template activity of both nucleolar and extranucleolar chromatin from rat prostate at various times after the onset of androgenic stimulation. A small activation of chromatin in both of these nuclear fractions was detected with *E. coli* RNA polymerase within 1 h of the administration of testosterone *in vivo*. For the reasons already stated, however, it would have been preferable to have used highly purified eukaryotic RNA polymerase in this study (MAINWARING et al., 1971). The low efficacy of bacterial polymerase to transcribe mammalian chromatin, particularly from the standpoint of detecting the localised and discrete changes evoked by hormones, has been forcibly demonstrated in the recent work of JACOB et al. (1975). In view of methods for preparing purified eukaryotic RNA polymerase B in high yields (JENDRISAK and BURGESS, 1975) this must be the enzyme of choice in future measurements of chromatin template activity.

Creditable attempts have been made to simulate the enhancement of RNA synthesis by steroid receptor complexes in reconstituted, cell-free systems. Prompted by earlier work on the activation of plant chromatin by plant auxins, *in vitro* (MATTHYSE, 1970; MATTHYSE and ABRAMS, 1970), receptor complexes for oestradiol-17β (MOHLA et al., 1972), 5α-dihydrotestosterone (DAVIES and GRIFFITHS, 1973; 1974; 1975), and testosterone (JACOB et al., 1975) are able to stimulate RNA synthesis with nuclei or chromatin as the source of template DNA *in vitro*. One objection to some of these studies was that the effects elicited by the steroid hormone were indirect or nonspecific, because the nuclear fractions were incubated with crude cytoplasmic fractions. Accordingly, the changes in template activity could be due to the hormone activating a nuclease and introducing nicks or other changes in DNA which were reflected in more efficient transcription. To circumvent problems of this nature, MAINWARING and JONES (1975) incubated chromatin with partially purified androgen receptor complex under conditions promoting the attachment of the complex to the acceptor sites. The chromatin was then sedimented by centrifugation at 10,000 *g*, prior to measurement of template activity with homologous (rat prostate) form B RNA polymerase under conditions where the amount of enzyme was not rate-limiting. Our results are presented in Table 11. The enhancement of RNA synthesis *in vitro* mandatorily required the receptor protein and 5α-dihydrotestosterone; in addition, prostate chromatin was more readily activated than spleen chromatin. Presumably, this was a reflection of their different content of acceptor sites (see section III.1.g). SBG could not mimic the role of the androgen receptor protein, for although the addition of human serum did stimulate RNA synthesis, this was not enhanced by the presence of 5α-dihydrotestosterone. The reason for the stimulatory effect of serum remains obscure. JACOB et al. (1975) have recently conducted similar experiments with the androgen receptor complex from mouse kidney. Two very important facts emerged from this elegant study; the activation of template activity by the testosterone-receptor protein complex showed clear evidence of saturation at a high input of cytoplasmic receptor, and the stimulation of transcription could be measured by homologous kidney RNA polymerase (both forms A and B) but not by heterologous bacterial enzyme (JACOB et al., 1975).

Collective evidence from several laboratories suggests that the advent of steroid-receptor complexes into chromatin can promote an increase in template activity. The precise mechanism of this activation remains to be elucidated. It is also essential that the effects observed to date with impure preparations of receptor complex should be confirmed with preparations of the highest possible purity. More sophisticated methods of analysis must also be employed for the characterisation of the RNA synthesized in cell-free conditions; nucleic acid hybridisation is an obvious candidate for future studies (for example, see REEDER, 1973; STEGGLES et al., 1974).

Table 11. Stimulation of the transcription of chromatin by receptor protein-5α-dihydrotestosterone complex *in vitro*.

Chromatin was prepared in manner to deplete endogenous RNA polymerase. Cytoplasmic extracts were prepared in presence (+) or absence (−) or 5×10^{-8} M 5α-dihydrotestosterone; receptor complex was selectively precipitated at 33% saturation with respect to $(NH_4)_2SO_4$. Chromatin was incubated with cytoplasmic preparations, collected by sedimentation at 10,000*g* and resuspended in medium suitable for assay of RNA synthesis in presence of partially purified prostate form B RNA polymerase. Data are taken from MAINWARING and JONES (1975) J. Steroid Biochem. **6**, 475.

Chromatin	Cytoplasmic preparation	5α-Dihydro-testosterone	RNA synthesis (c.p.m. H CTP incorporated/100μg of chromatin DNA)
Prostate (− enzyme)	None (controls)		160 ± 20
(+ enzyme)			660 ± 40
Spleen (− enzyme)			120 ± 20
(+ enzyme)			260 ± 60
Prostate	Prostate	−	640 ± 20
		+	1060 ± 20
Prostate	Spleen	−	280 ± 20
		+	260 ± 20
Spleen	Prostate	−	280 ± 20
		+	300 ± 40
Spleen	Spleen	−	220 ± 40
		+	200 ± 80

4. Conclusions

Over recent years, very considerable progress has been made in the elucidation of the biochemical events necessary for initiating androgenic responses. The high-affinity binding mechanism in rat prostate is now understood reasonably well, but the entry and release of steroid hormones from androgen target cells remain to be established in detail. The current model for the mechanism of action of androgens places critical importance on the nuclear retention of testosterone and its active metabolites, resulting in discrete changes in the transcription of the genetic apparatus. Evidence supporting this model has recently been achieved in reconstituted, cell-free systems. Alternative schemes for steroid hormone action, based on the association of hormones with lysosomes (SZEGO, 1974), do not seem appropriate for the mechanism of action of androgens (SEPSENWOL and HECHTER, 1976).

Current concepts derived from the widely favoured model system, rat ventral prostate gland, seem applicable to many other androgen target cells with the proviso that the active metabolite varies from one system to another. Developmental aspects of the mechanism of action of androgens must also receive wider attention in the future, especially in foetal and neonatal animals. Certain target cells (especially those in the brain) contain a spectrum of specific receptor proteins and their separation and characterisation is essential to our further understanding of the hormonal responses induced in these target sites. Investigators must also bear in mind that the significance of binding mechanisms must ultimately be reflected in dose-response relationships; evidence of this type is lacking at present. In addition, important biochemical responses must occur within the physiological concentration range of testosterone and its metabolites. In the rat, the plasma concentrations of androgens are known (GHANADIAN et al., 1975) and the concentrations of testosterone and 5α-dihydrotestosterone in many target cells have been determined (ROBEL et al., 1973).

The current mechanism of action of androgens is not readily applicable to the anabolic function of androgens in muscle; it remains a distinct possibility that an entirely different mechanism is in operation in muscle. Many aspects of a review by KOCHAKIAN (1975) support this contention. Similar reservations apply to the status of the binding of androgens in liver. In the experience of many investigators (ANDERSON and LIAO, 1968; MAINWARING, 1969 b; AHMED, 1971), rat liver is a nontarget organ for androgens; on the other hand, MILIN and ROY (1973) have suggested that an androgen receptor system is present in liver. By present standards, however, the evidence in this latter paper is not entirely convincing.

The solution of two problems is particularly pressing; the extensive purification of androgen receptor complexes and the investigation of the molecular basis for the stringent tissue specificity of androgenic responses.

IV. Early Events Stimulated by Androgens

Compared with experimental systems for other classes of steroid hormones, the responses mediated by androgens in their target cells are sluggish and take considerable time to attain their maxima once hormonal stimulation is begun. In animals depleted totally of androgens, say 3 or 7 days after bilateral orchidectomy, circulating concentrations of testosterone must be restored for 16–48 h before the majority of the biochemical and morphologic responses are fully expressed. The nature of these changes is the basis of this chapter. For the main part, the rat ventral prostate gland has been the experimental system of choice but its limitations in future work will be discussed.

1. Early Biochemical Changes Stimulated by Androgens

A survey of some of the biochemical changes promoted by androgens is presented in Table 12. The literature available on this subject is prodigious and the contents of Table 12 were selected to raise general points of importance only. In general terms, all the intracellular constituents of male accessory sexual glands are maintained by processes with early responses to androgens but taking considerable further time to reach their maxima.

From the historical standpoint, the identification of citric acid as a specific marker for androgenic stimulation was a turning point in studies on the mechanism of action of androgens at the biochemical rather than the morphologic level. The classical studies by Mann and his collaborators indicated the extreme specificity of androgenic responses; citric acid was a unique constituent of rat prostate (HUMPHREY and MANN, 1949; MANN and PARSONS, 1950) but was replaced by fructose in the seminal vesicle and prostate of many other species (SAMUELS et al., 1962). In all instances, the specific markers of fructose or citric acid were maintained only when the androgenic status of the animals was fully conserved. The Florence test for seminal fluid, widely used in forensic medicine, depends on the high concentration of polyamines in the secretion of the male accessory sexual glands. In most species, the prostatic secretion is among the richest sources of polyamines (MANN, 1964) and the intracellular concentrations of spermine and spermidine, together with the enzymes required for their synthesis, ornithine decarboxylase, and S-adenosylmethionine decarboxylase, are stringently regulated by androgens (PEGG and WILLIAMS-ASHMAN, 1969; PEGG et al., 1970). The influence of polyamines on the rates of growth of bacteria and eukaryotic cells is very considerable (for reviews, see PEGG et al., 1970 and RUSSELL, 1973), for in addition to stabilising DNA and ribosomes, they accelerate nucleic acid and protein synthesis. In many systems, evidence favouring a close coordination between polyamine synthesis and ribosomal RNA synthesis is compelling but a causal relationship remains to be clearly established. The best indication of this functional interdependence is provided by the anucleolate mutant of the South African clawed toad,[2] *Xenopus laevis,* where the synthesis of both ribosomal RNA and polyamines is virtually absent (RUSSELL, 1971). The precise reason for this functional correlation during RNA synthesis is uncertain but the cationic polyamines may neutralise the excess of negative charges on newly synthesized ribosomal RNA and play an important part in the assembly and stabilisation of the functional polyribosomes required for protein synthesis.

[2] Although widely described as a toad, it is more taxonomically correct to be classified as a frog.

Table 12. Early biochemical events stimulated by androgens.

The numbers of the enzymes, given in brackets, are those recommended by the International Union of Biochemistry (1972)

Type of process	Reference
(a) Induction of enzymes	
Cytochrome oxidase (1.9.3.1)	DAVIS et al. (1949)
Malate dehydrogenase (1.1.1.82)	WILLIAMS-ASHMAN (1954)
Citrate-condensing enzyme (4.1.3.7)	WILLIAMS-ASHMAN and BANKS (1954)
Glycosidases (eg. 3.2.1.31) [a]	CONCHIE and FINDLAY (1959)
Aldolase (4.1.2.13)	BUTLER and SCHADE (1958)
Aldose- and ketose-reductase (eg. 1.1.1.21)	SAMUELS et al. (1962)
Adenosinetriphosphatase (3.6.1.3) [b]	AHMED and WILLIAMS-ASHMAN (1969)
RNA polymerase (2.7.7.6)	MAINWARING et al. (1971)
(b) Synthesis of constituents of low molecular weight	
Citric acid	HUMPHREY and MANN (1949); MANN and PARSONS (1950)
Polyamines (spermidine and spermine) [c]	PEGG and WILLIAMS-ASHMAN (1969); PEGG, et al. (1970)
(c) Synthesis of macromolecular constituents	
Polyribosomes	MAINWARING and WILCE (1973)
Messenger RNA [d]	MAINWARING et al. (1974 c)
Endoplasmic reticulum (membranes)	MAINWARING and WILCE (1972)
Nuclear proteins (nonhistones)	CHUNG and COFFEY (1971 a, b)
Nuclear membranes	CHUNG and COFFEY (1971 a)

[a] a particularly good marker is β-glucuronidase; [b] microsomal enzyme, requiring both Na^+ and K^+ ions; [c] quantitative measurements on the polyamines and enzymes engaged in their synthesis, eg. ornithine decarboxylase; [d] poly(A)-rich messenger RNA, only.

An extremely wide spectrum of enzyme activities is induced by androgens in the prostate and other male accessory sexual glands. One important feature of the androgen-sensitive enzymes listed in Table 12 is that they represent activities in all the subcellular fractions of the prostate, ranging from nuclei (RNA polymerase; MAINWARING et al., 1971), mitochondria (cytochrome oxidase; DAVIS et al., 1949), lysosomes (β-glucuronidase; CONCHIE and FINDLAY, 1959), microsomes (adenosinetriphosphatase; AHMED and WILLIAMS-ASHMAN, 1969) to the cell soluble fraction (aldolase; BUTLER and SCHADE (1958). This indicates the extremely wide scope of androgen-promoted phenomena, encompassing all intracellular compartments of target cells. The molecular basis for enzyme induction by androgens is discussed in greater detail in the next section (Chapter IV.2). As one would expect, enzymes engaged in the synthesis of citric acid and fructose are extremely sensitive indicators of androgenic stimulation (WILLIAMS-ASHMAN and BANKS, 1954; SAMUELS et al., 1962). The microsomal Na^+-K^+-dependent adenosinetriphosphatase was originally believed to be activated directly by testosterone and 5α-dihydrotestosterone, even under cell-free conditions, *in vitro* (FARNSWORTH, 1968); however, this effect could not be repeated by other investigators (AHMED and WILLIAMS-ASHMAN, 1969).

The enhancement of protein synthesis in rat prostate by androgens has been extensively studied (for an excellent review of early work, see WILLIAMS-ASHMAN and REDDI, 1971). Cell-free systems for monitoring the incorporation of radioactive amino acids into polypeptide linkage, *in vitro,* were the favoured experimental approaches, both in cytoplasmic ribosomes and microsomes (LIAO and WILLIAMS-ASHMAN, 1962; MANGAN et al., 1967) and mitochondria (PEGG and WILLIAMS-ASHMAN, 1968). The experience of all investigators was that the stimulation of protein synthesis was a relatively slow androgenic response, requiring at least 24 h to reach its maximum. Much of the early work centred on the measurement of protein synthesis from which the membranes of the endoplasmic reticulum had been removed by surface-active detergents, such as sodium deoxycholate and Triton X-100. As reviewed by CAMPBELL and LAWFORD (1967), however, the cytoplasmic membranes control the release of proteins for "export" from secretory cells and all male accessory sexual glands produce copious secretions, rich in proteins. Prompted by these observations, MAINWARING and WILCE (1972) studied the influence of androgens on the synthesis and turnover of membranes of the endoplasmic reticulum of rat prostate in detail. By using centrifugation in discontinuous sucrose gradients, four discrete fractions were resolved from prostate microsomes; smooth membranes (free of ribosomal particles), light and heavy rough membranes (with low and high contents of ribosomes, respectively), and free ribosomal particles (free of membranes). From studies performed both *in vivo* and *in vitro,* the majority of the synthesis of proteins in rat prostate took place in the heavy rough fraction and this was subject to extreme androgenic regulation. Furthermore, the incorporation of [^{14}C] choline or [^{32}P]H_3PO_4 into membrane phospholipids was controlled by androgens and the quantitative assay of all membrane constituents (phospholipids, proteins, and ribosomal RNA) confirmed that the entire structure of the endoplasmic reticulum was maintained only by high circulating concentrations of androgens. As would be expected of a secretory organ, therefore, the membranes of the rat prostate were a dynamic system with a rapid turnover and a high rate of renewal (MAINWARING and WILCE, 1972). This conclusion is in harmony with the findings of DALLNER et al. (1966) and ARIAS et al. (1969) in other secretory cells. From time-course experiments after the injection of testosterone into castrated rats, an interesting difference was found in the relative rates of synthesis of the components of the endoplasmic reticulum. Synthesis of ribosomal RNA was stimulated maximally only 2 h after the onset of androgenic stimulation (MAINWARING et al., 1971), yet the membranes and functional microsomes were present in appreciable amounts only some 16–24 h later. The transport, processing and assembly of newly fabricated ribosomal particles in the cytoplasm is a complex process, depending on the provision of polyamines, structural ribosomal proteins, and divalent cations. Little is known about the accumulation of Mg^{2+} ions in the prostate but the synthesis of all the remaining components is strictly controlled by androgens (PEGG et al., 1970; MAINWARING and WILCE, 1972). Our findings are totally consistent with the general concept proposed by TATA (1967) that the proliferation of cytoplasmic membranes accompanies the hormone induced synthesis of ribosomal particles.

While advancing knowledge considerably, most studies to date have suffered from three shortcomings. First, protein synthesis has been studied in rather general terms; what is needed now is a more concerted effort on the synthesis of specific and well-characterised proteins. Second, androgenic control is attributed by nearly all investigators to the provision of active, functional microsomes and a wealth of other translational control mechanisms has been sadly overlooked. The studies begun by LIANG and LIAO (1975 a, b) must be extended. Third, the steroid- and tissue-specificity of protein synthesis is deploringly neglected in many studies. This precludes an assessment of the involvement of the receptor mechanisms in initiating many of these responses. The most informative appraisal of the effects of androgens on translational processes is the study by ICHII et al. (1974). Their important findings are as follows. (1) Androgens regulate the concentration of poly(A)-rich RNA particles in prostate cytoplasm rather than the nucleus, indicating a considerable measure of hormonal regulation in the processing and modification of cytoplasmic RNA. (2) The complement of active polyribosomes is modulated by testosterone, *in vivo.* (3) Androgens control the activity of initiation rather than elongation factors for protein synthesis. It is to be hoped that this important work will be extended in the future.

Rather more sophisticated studies on the androgenic control of the synthesis of nuclear proteins were conducted by CHUNG and COFFEY (1971 a, b). They demonstrated that certain classes of nuclear-associated proteins in rat prostate required androgens for their maintenance, particularly in the acidic or nonhistone protein fraction. Histones were less sensitive to androgenic modulation, except histone F1, which was severely depleted in castrated animals yet rapidly restored after the administration of testosterone, *in vivo* (CHUNG and COFFEY, 1971 a). This is an important observation because histone F1 is believed to be implicated in the onset of DNA synthesis (BRADBURY et al., 1973; 1974 b); this androgenic response is discussed in detail in Chapter V. While the prostate nonhistone proteins were regulated by androgens both qualitatively and quantitatively, they were synthesized prior to the onset of DNA replication (CHUNG and COFFEY, 1971 b) and it was suggested that these macromolecular events were not directly coupled. A corollary of this work is that the structural genes coding for nuclear proteins are freely accessible and therefore active without the extensive opening up of the structure of chromosomes which must precede mitosis.

As so well reviewed by WILLIAMS-ASHMAN and REDDI (1971) and VILLEE et al. (1975), the androgenic control of RNA synthesis has been almost exclusively studied from the standpoint of the supply of ribosomal RNA or nonribosomal RNA whose form and function was not established with any certainty. As a typical representative, MAINWARING et al. (1971) demonstrated that the administration of androgens to castrated rats rapidly stimulated the nucleolar RNA polymerase (form A, for ribosomal RNA synthesis) and a slower but equally impressive stimulation of the extra-nucleolar RNA polymerase (form B, for nonribosomal and messenger-like RNA synthesis). Similar changes have been noted recently by JACOB et al. (1975) in mouse kidney, but in neither case was the product of the form B enzyme rigorously characterised. Until the work by MAINWARING and WILCE (1973) and MAINWARING et al. (1974 c) was published, little was known of the androgenic regulation of messenger RNA synthesis. MAINWARING and WILCE (1973) first established that androgens completely regulated the synthesis and assembly of rat prostate polyribosomes in a strictly tissue- and steroid-specific manner. In castrated animals polyribosomes were severely depleted, possessed little synthetic activity and were mainly present only as ribosome dimers, $S_{20,\omega}$ 124S. Within 24 h of androgenic stimulation *in vivo,* the content of polyribosomes was fully restored and the newly synthesized RNA complexes were conspicuously active in protein synthesis. As judged by ultracentrifugation analysis, these polyribosome complexes contained at least ribosome hexamers, $S_{20,w}$ 240S and greater. On the basis of these findings, it was concluded by MAINWARING and WILCE (1973) that androgens regulated the synthesis of messenger RNA in the rat prostate and this was convincingly confirmed by MAINWARING et al. (1974c). A fraction of prostate RNA was isolated which fulfilled all the exacting criteria demanded of a putative messenger RNA. First, this RNA fraction was rapidly labelled with [^{3}H] uridine and was extremely sensitive to the metabolic inhibitor, actinomycin D. Second, the RNA formed stable duplexes with immobilised nucleic acids, including oligo(dT)-cellulose (AVIV and LEDER, 1972) and poly(U)-glass fibre discs (SHELDON et al., 1972). These matrices provided ready means for isolating the poly(A)-enriched RNA fraction, particularly from the RNA contained in prostate polyribosome complexes (200S and greater). Third, this prostate RNA fraction was selectively labelled with the RNA precursor, [^{14}C] 5′-fluoro-orotic acid, which is essentially incorporated into messenger RNA only (WILKINSON et al., 1971). Fourth, the RNA fraction sedimented as a broad, heterogeneous peak in sucrose gradients, sedimentation coefficient 6 to 15S, and was the only prostate RNA fraction to stimulate protein synthesis in a cell-free system derived from Krebs II ascites tumour cells and depleted in endogenous messenger RNA (MATHEWS and KORNER, 1970). Taken collectively, the results of MAINWARING et al. (1974c) provide irrefutable evidence that androgens regulate the synthesis of messenger RNA in rat prostate and this important conclusion is supported by the independent study of ICHII et al. (1974). Studies on other androgen target cells have not been published.

It should be stressed that all the studies on the messenger RNA of androgen target cells thus far have been directed toward species conspicuously enriched with poly(A)-sequences at the 3′-terminus, largely because the technology for the isolation and assay of this type of messenger RNA is well advanced. However, as reviewed by DARNELL et al. (1973) and LEVIN (1975), the eukaryotic messenger RNA capable of participating in cytoplasmic protein synthesis is the end product of an intricate processing mechanism beginning with heterogeneous nuclear RNA. Consequently, the messenger RNA of mammalian cells is exceedingly complex; some species do not contain poly(A)-sequences (ADESNIK and DARNELL, 1972; NEMER et al., 1975), some heterogeneous nuclear RNA is notably rich in poly(U)-sequences rather than poly(A)-sequences (MOLLOY et al., 1972) and some species contain long tracts of poly(C)-sequences (PORTER et al., 1975). On this evidence, the heterogeneous nuclear RNA and cytoplasmic messenger RNA are complicated mixtures of polyribonucleotides and androgenic regulation of other than poly(A)-rich representatives of these fractions can safely be predicted in future investigations. In addition, the processing of heterogeneous nuclear RNA may provide another important mode of androgenic regulation.

Prompted by earlier work on the oestrogen-primed uterus of immature rats (SEGAL et al., 1965) a novel approach to the androgenic regulation of protein synthesis has been pursued by VILLEE et al. (1975). Testosterone or other steroids are injected in castrated animals, with experimental controls receiving injections of saline only and at various times thereafter, total cellular RNA or poly(A)-rich RNA is extracted from androgen target tissues, such as seminal vesicle. The RNA from the stimulated animals is then instilled with [^{3}H] leucine into one seminal vesicle of a castrated rat whereas the contralateral seminal vesicle received the radioactive amino acid and RNA extracted from the controls; after 4 h, the radioactivity incorporated into protein was determined. The instillation of RNA from androgen-primed animals provoked a significant enhancement in protein synthesis. In a sense, this experimental success is remarkable because prostate poly(A)-rich RNA is acutely sensitive to degradation by ribonuclease (MAINWARING et al., 1974 c) and one would expect the instilled RNA to be rapidly inactivated. We really know very little about the entry of macromolecules into intact cells and it should be demonstrated unequivocally that instilled RNA, previously labelled to a very high specific radioactivity with say, ^{32}P, really enters the seminal vesicles intact. Otherwise, the possibility that the instilled RNA is serving merely to protect endogenous messenger RNA cannot be discounted; this is an important point in androgenic responses since they proceed almost exclusively by amplification mechanisms. By definition, these involve quantitative rather than qualitative modulations of RNA synthesis.

The control of physiological rather than biochemical processes by androgens was not included in Table 12 because their precise basis in molecular terms is unclear. Studies by SHIMAZAKI et al. (1973) indicate that the composition of the pools of free amino acids in rat prostate is not sensitive to androgens. By contrast, the uptake of glucose (HÄRKÖNEN et al., 1975) and nucleoside precursors of RNA (HONMA and NOUMURA, 1975) are both controlled by testosterone or its metabolites. The latter study is important because increases in rapidly labelled, messenger-like RNA after androgenic stimulation could be attributed to changes in pool size alone. This stricture may apply to certain studies but not to that of MAINWARING et al. (1974 c) using [^{14}C] 5′-fluoro-orotic acid as precursor, because this has no strict intracellular counterpart. Based on the original observations of MAWSON and FISHER (1952), zinc (or more accurately, Zn^{2+} ions) is actively concentrated in the prostate glands of many species, including man and is a unique constituent of prostatic secretions. The amounts transcend those required for metallo-enzyme prosthetic groups alone. REED and STITCH (1973) have identified a zinc-binding protein in human prostate gland distinct from the androgen-binding proteins. Of extreme interest was the observation that cadmium could significantly displace the binding of zinc, because cadmium has deleterious and even carcinogenic effects on the reproductive tract (PÂRÍZEK, 1957). The precise role of zinc in male accessory sexual glands is not clear but it is located in secretory granules (MAQUINAY et al., 1963), endoplasmic reticulum and nucleolar apparatus (CHANDLER et al., 1975).

Despite the slowness of all these biochemical responses, they are initially set in train by the androgen receptor system. This correlation cannot be drawn in all instances, but as emphasised in Chapter I.3a, all of the changes reported from this laboratory are sensitive to the antiandrogens, BOMT and cyproterone acetate (MAINWARING and WILCE, 1972; MANGAN and MAINWARING, 1972; MAINWARING et al., 1974 b, c).

2. A More Detailed Enquiry into Enzyme Induction by Androgens: Aldolase in Rat Prostate

Since the induction of enzymic activity is such a prominent feature of the early androgenic responses (see Table 10), this topic demands closer consideration. Assuming that the enzyme substrates are not limiting, then the stimulation of these enzymes could be attributed to many factors: increased synthesis of enzyme moieties de novo, the enhanced assembly of preformed, stable subunits of enzymes, accelerated production of rate-limiting prosthetic groups or a reduced rate of enzyme degradation. Evidence is not available on all these points, but the detailed study by MAINWARING et al. (1974 b) on the synthesis of rat prostate aldolase is germane to enzyme induction in the widest context.

Three considerations favoured the selection of aldolase as the model enzyme for this study. First, aldolase activity is regulated by androgens (BUTLER and SCHADE, 1958). Second, it has distinctive physicochemical properties which may be profitably used for the purpose of identification. Third, it may be readily isolated by two cycles of affinity chromatography on a matrix of phosphocellulose with a NaCl gradient as eluent for the first column and elution with substrate (fructose-1,6-diphosphate) for the second column (GRACY et al., 1970). The aldolase was finally crystallised by dialysis against gradually increasing concentrations of $(NH_4)_2SO_4$ (GRACY et al., 1969). Antibodies were then raised in rabbits against crystalline rat prostate aldolase, with control serum being taken from animals that had received injections of the adjuvant vehicle only. This provided a specific and extremely sensitive means for monitoring the synthesis of aldolase by immunological precipitation. This glycolytic enzyme is represented in eukaryotes by a spectrum of isoenzymes, which are classified in relation to three basic forms A, B, and C, as typified by the principal enzyme in muscle, liver, and brain revealed by electrophoresis (RUTTER, 1964). The purified aldolase of rat prostate was the muscle or type A enzyme. This need not mean that the enzyme was of stromal rather than epithelial origin, because the type A enzyme is present in many tissues other than muscle; indeed the muscle-type enzyme is present in hepatomas rather than the type B form of normal liver (GRACY et al., 1970; IKEHARA et al., 1970). This point is important because the androgen-responsive cells in rat prostate are predominantly in the glandular epithelium rather than the underlying muscular stroma (TVETER and ATTRAMADAL, 1969).

The evidence accumulated by MAINWARING et al. (1974 b) overwhelmingly supports the contention that androgenic induction is primarily explained by the synthesis de novo of the messenger RNA coding specifically for the enzyme polypeptide chain. The grounds in support are as follows. (1) The enhancement of aldolase activity in castrated animals was prevented by the concomitant injection of actinomycin D or cycloheximide with testosterone, *in vivo*. This suggests that continued RNA and protein synthesis is an obligatory feature of the induction process. (2) Using both *in vivo* and *in vitro* methods, the induction of the messenger RNA for prostate aldolase was regulated by androgens in a stringently tissue- and species-specific manner. In the approach *in vivo*, [^{35}S] methionine was injected directly into the prostate glands of castrated animals that had been maintained in various hormonal regimens. Labelled aldolase was then selectively precipitated from isolated polyribosomes with the anti-aldolase serum. Aldolase was induced only by testosterone, the response being maximal after 16–24 h of androgenic stimulation; corticosterone and oestradiol-17β could not mimic this effect. Heavy prostate polyribosomes, of sedimentation coefficient 200S and above, were responsible for the synthesis of aldolase, this complement of ribosome hexamers being almost absent in castrated animals. In the procedure *in vitro*, the poly(A)-rich messenger RNA was isolated from the prostate and liver of castrated animals at various times after the administration of testosterone and added to a protein-synthesizing system with [^{3}H] phenylalanine as tracer. By immunoprecipitation analysis, labelled aldolase was detected among the polypeptides made in the presence of the messenger RNA of prostate and liver, but only that in prostate was stimulated by testosterone. The aldolase message was present in maximal amounts after 8–16 h of androgenic stimulation and then declined. (3) The results strongly suggest that the androgen receptor system triggers the synthesis of this specific messenger RNA and hence enzymic induction is best explained by transcriptional changes. However, the difference in time between the appearance of the messenger RNA and the completed enzyme polypeptide chain, distinguished by the procedures *in vitro* as against *in vivo*, suggest that androgens may have some measure of control over translational events as well. This could be in the transport or processing of the message and the provision of initiation factors (see also ICHII et al., 1974). (4) The aldolase messenger RNA was detectable even in castrated rats prompting the conclusion that androgenic responses, as typified by enzyme inductions, proceed by amplification mechanisms.

There is no reason *a priori* why the findings on aldolase synthesis are not applicable to all the enzymes subject to androgenic control. The present evidence throws no light on other aspects of induction, such as the stabilisation of messenger RNA, the curtailment of enzyme degradation, or the accelerated release of nascent enzyme chains from polyribosomes. Nevertheless, the molecular basis of enzyme induction is now understood in considerable detail.

3. Experimental Models for Studying Tissue-Specific Responses

Transcending most other aspects of androgenic responses is the satisfactory solution of tissue specificity. Something of an impasse exists because current methods for analysing the binding of androgens have provided few insights into the underlying maintenance of tissue specificity. An alternative approach is to monitor the androgenic regulation of a tissue-specific protein and its specific messenger RNA, but suitable systems have only recently been developed for such work. Two factors militated against an extension of studies on the induction of rat prostate aldolase; first, the prostate (type A) aldolase is not tissue-specific and second, it represents but a minor fraction of the total protein of soluble extracts of the prostate, approximately 0.1% only. Two other systems, however, offer hope for future investigations.

a) Rat Seminal Vesicle

In male rodents, a major part of the seminal plasma is derived from the secretion of the seminal vesicle (MANN, 1964) and the growth of this accessory sexual gland requires the continuous presence of androgens (MOORE et al., 1930). The proteinaceous constituents of the seminal secretion have been widely studied in the past (BALLARD and WILLIAMS-ASHMAN, 1964; MANYAI, 1964; NOTIDES and WILLIAMS-ASHMAN, 1967) but have received little attention in recent years. Seminal secretion is rapidly coagulated by the enzyme, coagulase, secreted by the coagulating gland adjacent to the seminal vesicle and after ejaculation, the proteins in the mixed seminal plasma are involved in the formation of the vaginal plug (BLANDEAU, 1945; PRICE and WILLIAMS-ASHMAN, 1961; JOSHI et al., 1972). The basic proteins described by MANYAI (1964) and NOTIDES and WILLIAMS-ASHMAN (1967) are generally believed to provide the structural basis for the vaginal plug and are accordingly termed "clotting proteins."

A continuation of these studies by HIGGINS et al. (1976) indicates that certain proteins of the seminal vesicle warrant study from the standpoint of tissue specificity. As illustrated in Figure 9, certain basic proteins of rat seminal vesicle are regulated by androgens in a remarkably stringent manner. These proteins may only be detected in rat seminal vesicle and no other secondary sexual gland. In addition, their synthesis in castrated animals may only be stimulated by androgens and the specificity of the induction is a perfect reflection of the binding affinities of the androgen receptor mechanism in the seminal vesicle (MAINWARING and MANGAN, 1973). Accordingly, the concomitant administration of cyproterone acetate with testosterone suppresses the induction of the proteins *in vivo* whereas corticosterone does not.

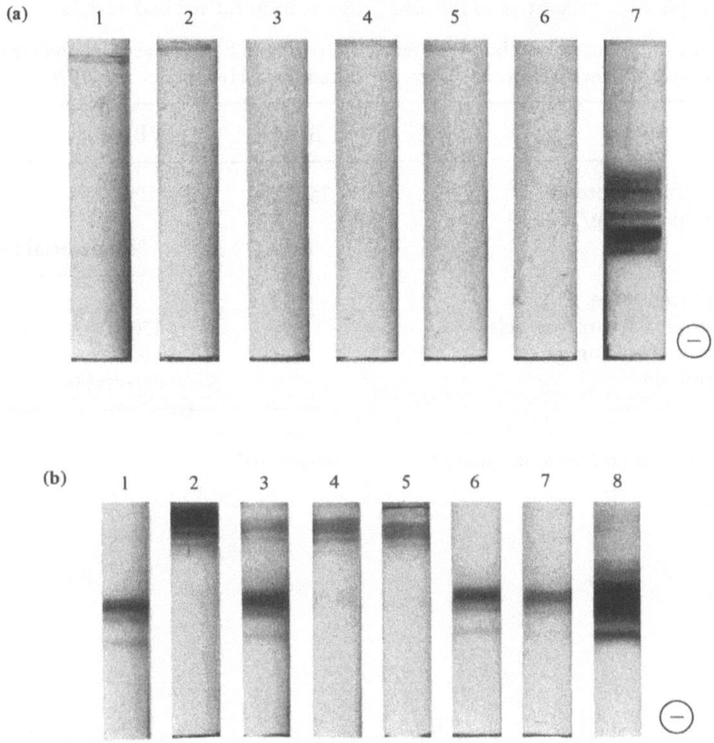

Fig. 9 a–b. Basic proteins of rat seminal vesicle. (a) Proteins were extracted from tissues of normal, noncastrated rats, and those soluble in 2M NaCl at 100° C were analysed in polyacrylamide gels at pH 4.2 in 3M urea. Direction of cathode is indicated. Tissues: 1 liver; 2 spleen; 3 testis; 4 preputial gland; 5 ventral prostate gland; 6 caput epididymis; 7 seminal vesicle
(b) Castrated rats were injected with various steroids (1 mg; unless stated otherwise) and basic proteins of seminal vesicle were analyzed as in (a). Steroid treatments: 1 testosterone alone; 2 cyproterone acetate alone, 10 mg; 3 testosterone plus cyproterone acetate; 4 oestradiol-17β alone; 5 corticosterone alone; 6 corticosterone plus testosterone; 7 testosterone, 0.5 mg; 8 testosterone, 5.0 mg. Differences beween gels 1, 3, and 6 were even more marked when scanned in a microdensitometer. Gels are reproduced from HIGGINS et al. (1976) Biochem. J. 158, 271

By using gel-exclusion chromatography on Sephadex G-200, followed by preparative polyacrylamide electrophoresis under the denaturing conditions imposed by the detergent, sodium dodecyl sulphate, two proteins of rat seminal vesicle have been purified to apparent homogeneity. These differ only in their electrophoretic mobility and their properties are summarised in Table 13. The proteins have been designated as SVBP/F and SVBP/S on the basis of their fast and slow migration in polyacrylamide gels and their important features are as follows. (a) Together they constitute a major proportion of the seminal vesicle proteins, to some 25–30%. (2) Both SVBP/F and SVBP/S are glycoproteins and may be detected by their positive reaction to the periodic acid-Schiff base test (p-rosaniline). (3) Both are extremely basic proteins,

Table 13. Properties of purified proteins from rat seminal vesicle.

SVBP is mixture of two closely-related proteins; SVBP/F and SVBP/S are fast and slow forms resolved by electrophoresis. Data are taken from HIGGINS et al. (1976).

Property	SVBP	SVBP/F	SVBP/S
Content of vesicular proteins	25–30%	10%	20%
Carbohydrate content (g/100g) [a]			
D-Mannose	0.31	Not determined	
D-Galactose	0.37		
N-Acetyl-D-glucosamine	0.53		
Molecular weight (denaturing gels)	—	17000	18500
Isoelectric point (focusing gels)	—	9.7	9.7
N-Terminal residue [b]		None detectable	
Amino acid analysis (n moles/m mole) [c]	not determined		
Cysteine		11	4
Aspartic acid		97	60
Threonine		35	17
Serine		147	205
Glutamic acid		152	177
Proline		35	31
Glycine		81	52
Alanine		60	56
Valine		35	46
Methionine		34	18
Isoleucine		33	32
Leucine		34	34
Tyrosine		9	28
Phenylalanine		48	36
Histidine		14	14
Lysine		93	105
Arginine		84	86

[a] N-acetyl neuraminic acid and N-acetyl-D-galactosamine are absent; [b] by both phenylisothiocyanate and fluorodinitrobenzene; [c] no determination of amides or tryptophan.

but their N-terminal groups could not be located even with [^{3}H]-1-fluoro-2,4-dinitrobenzene. (4) In keeping with their basic pI, they are both exceedingly rich in lysine and arginine but low in hydrophobic amino acids. Presumably the aspartic and glutamic acids are predominantly present as their neutral amides. (5) SVBP/F and SVBP/S undergo classical induction and deinduction with respect to androgens. After castration, the two proteins almost disappear beyond the limits of detection in 7–10 days; when the androgenic status of castrated animals is restored, the basic proteins are synthesized de novo to their levels in noncastrated animals in 2 or 3 days. Their regulation seems to be a good example of an amplification phenomenon.

Despite a superficial similarity in properties, these two proteins described by HIGGINS et al. (1976) are not necessarily identical to the clotting proteins described by others (MÁNYAI, 1964; NOTIDES and WILLIAMS-ASHMAN, 1964). There seem to be three points of distinction. First, other investigators have not described carbohydrates as integral components of their basic proteins. Second, there are differences in the amino acid composition of SVBP/F and SVBP/S as compared with other seminal proteins of high pI. In particular, the protein described by BALLARD and WILLIAMS-ASHMAN (1964) and NOTIDES and WILLIAMS-ASHMAN (1967) is devoid of tyrosine, proline, and cysteine, but these are present in SVBP/F and SVBP/S (see Table 13). MÁNYAI (1964) reported a lysine:arginine ratio of 4:1 for his clotting protein whereas that of the proteins isolated by HIGGINS et al. (1976) is nearer to 1:1. Third, SVBP/F and SVBP/S are not exquisitely sensitive to precipitation (or clotting) by coagulase; this was the characteristic feature of the protein described by NOTIDES and WILLIAMS-ASHMAN.

In summary, the biological function of the two basic proteins purified in this laboratory remains somewhat conjectural, but in terms of size, content, ease of identification, distinctive properties, and particularly their unique distribution, they are candidates for extensive studies on the tissue specificity of androgenic responses. It will be necessary to establish whether these proteins contain an identical, fundamental subunit, thus differing only in minor aspects of geometric assembly or quaternary structure. N-terminal analyses did not clarify this point but fingerprinting of tryptic digests may be informative. Our intention is to isolate the messenger RNA(s) coding for these proteins and to establish unequivocally whether the synthesis of this RNA fraction is indeed specific to the seminal vesicle. Such experiments would go a long way in explaining the molecular basis for tissue-specific responses.

b) Mouse Kidney

Without question, the most advanced enquiry into the tissue-specific induction of an enzyme by androgens is presented by β-glucuronidase in mouse kidney, as first described by FISHMAN (1951). This favourable situation may be attributed to the outstanding work of PAIGEN, GANSCHOW, and their co-workers on the molecular architecture of β-glucuronidase and to the availability of inbred and recombinant lines of mice bearing stable mutations in the genetic loci responsible for the synthesis of this hydrolase. The importance of these genetic variants has been emphasized by SWANK and BAILEY (1973).

Glucuronidase is somewhat unusual in that it is located in two subcellular fractions of mouse kidney, the lysosomes, and microsomes (PAIGEN, 1961 a; FISHMAN et al., 1969). The enzyme in both intracellular sites is identical in terms of immunological determinants and kinetic parameters (PAIGEN, 1961 a, b; LALLEY and SHOWS, 1974) and thus coded by the same structural gene. Despite this structural homology, there are mutants in which only the lysosomal enzyme is present (GANSCHOW and PAIGEN, 1967, 1968). Elegant studies by SWANK and PAIGEN (1973) distinguished six forms of glucuronidase in mouse kidney, designated in the order of their decreasing electrophoretic mobility towards the anode in polyacrylamide gels. Together with their molecular weights and intracellular location, they were named as follows: L (lysosomes; 260,000), then M1, M2, M3, M4 (microsomes; 310,000–470,000), and finally X (microsomes; 260,000), of lower mobility than form L but the species of β-glucuronidase induced by androgens, *in vivo* (GANSCHOW and BUNKER, 1970). A fundamental unit, molecular weight approximately 70,000, provides the polypeptide chain for all the forms of kidney glucuronidase, these all having tetrameric structures in the functional moiety of the enzyme. Conversion of the various M forms into form X can be achieved by treatment with trypsin or urea (SWANK and PAIGEN, 1973) and it is now clear that the microsomal or M forms contain increasing amounts of another protein, one chain in the M1 tetramer, two in the M2 tetramer and so on. This additional protein, called egasyn, has been purified to homogeneity (TOMINO and PAIGEN, 1973) and like the proteins associated with other enzymes (STRITTMATTER et al., 1972; SPATZ and STRITTMATTER, 1973), egasyn enables β-glucuronidase to bind to the membranes of the endoplasmic reticulum.

Once the basic architecture of the enzyme was fully established, it was then possible to delineate precisely the genes necessary for the expression and regulation of all forms of β-glucuronidase in mouse kidney. This is a truly remarkable achievement and a summary of the genetic elements involved in this induction process are summarised in Table 14. Up to six genes are involved and each controls discrete aspects of the induction process with respect to intracellular distribution, developmental changes, enzyme translocation, and sensitivity to hormones. These features are presented in the authoritative paper by PAIGEN et al. (1975) where they propose that the genes for β-glucuronidase fall into four categories. (1) The structural gene (*gus*) determines the polypeptide sequence of the monomer subunit of glucuronidase, including the so called recognition amino acid sequences which may control enzyme activity in a genetically determined manner. These recognition sequences are believed to be important in the modification of the enzyme subunit after its synthesis, by phosphorylation or other reactions in the kidney cytoplasm. The X gene is clearly needed for the synthesis of the androgen receptor protein, necessary for the transfer of testosterone or 5α-dihydrotestosterone in the "high information" or receptor-bound form into kidney chromatin. Both the *gus* and X genes are structural genes, regulating the synthesis of a messenger RNA for the enzyme and receptor proteins, respectively. There are three alleles at the *gus* locus. (2) The processing gene (*Eg*) controls the synthesis of the protein, egasyn, necessary for the intracellular localisation of glucuronidase and even, perhaps, its degradation. The *bg* locus represents a further processing gene, involved in the secretion of β-glucuronidase from the kidney into the general circulation. (3) The regulatory gene (*gur*) is responsible for controlling the rate of enzyme synthesis, notably in response to androgenic stimulation. Presumably, the androgen receptor complex binds at the *gur* locus. (4) The temporal gene (*gut*) determines the regulation of enzyme synthesis as demanded by the processes of growth and differentiation. The scheme proposed for the genetic regulation of β-glucuronidase is applicable in principle to many other target cells where enzyme induction is enhanced by androgens. In particular, it is most likely that the structural, regulatory, and temporal genes will be in close proximity within the genome, as evident here with chromosome 5, whereas the processing genes are likely to be linked at a different genetic locus, in this case chromosome 8. These concepts are embodied in Figure 10, taken from PAIGEN et al. (1975).

It was suggested earlier in this monograph (Chapter I.4.g) that under certain circumstances the androgenic induction of β-glucuronidase in mouse kidney could occur through purely cytoplasmic or translational mechanisms and this is counter to the present propositions of PAIGEN et al. (1975). However, the translational mechanism was proposed exclusively for the rapid but transient enhancement of enzyme activity

Table 14. Genetic loci involved in regulation of β-glucuronidase in mouse kidney.

Genetic locus	Gene linkage	Class of genes [a]	Strain of mice	Function	Reference
Gus	Chromosome 5	Structural gene	A/J (gus [a]) C 57 BL/6J (gus [b]) C 3 H/He (gus [h])	Synthesis of enzyme polypeptide monomer; gus [b] is standard allele, gus [a] produces enzyme of different electrophoretic mobility and gus [h] produces thermolabile enzyme	Law et al. (1952) Swank et al. (1973) Lally and Shows (1974)
Eg	Chromosome 8	Processing gene	Eg° mutation in YBR	Controls synthesis of egasyn, and thus distribution of enzyme between lysosomes and microsomes	Ganschow and Paigen (1967, 1968) Swank et al. (1973) Karl and Chapman (1974)
Gur	Chromosome 5; close to gus and recessive gene, buff, for coat colour	Regulatory gene	A and C 57 BL/6 bf/bf, their F1 and F2 hybrids and back-cross progeny	Controls induction of X-enzyme form by androgens; low and long lag induction in C 57 BL/6, but high and rapid induction in A. Acts in cis manner to gus and is expressed solely in mouse kidney	Swank et al. (1973) Paigen et al. (1975)
Gut	Chromosome 5; close to gus but in terminal chromosome segment	Temporal gene	C 3 H/HeHa, DBA/LiHa and their F1 hybrids	Controls appearance of glucuronidase during development, possibly in tissue-specific manner	Paigen (1961 b) Paigen et al. (1975)
Tfm	X (male) chromosome	Regulatory gene	Tfm	Controls male sexual differentiation and in particular, synthesis of specific androgen receptor in androgen target cells, including kidney	Lyon and Hawkes (1970) Dofuku et al. (1971) Bullock and Bardin (1972) Attardi and Ohno (1974)
bg	Chromosome 13	Processing gene (?)	—	Influences rate of secretion or release of glucuronidase	Quoted (unpublished work) in Paigen et al. (1975)

[a] classification from Paigen et al. (1975)

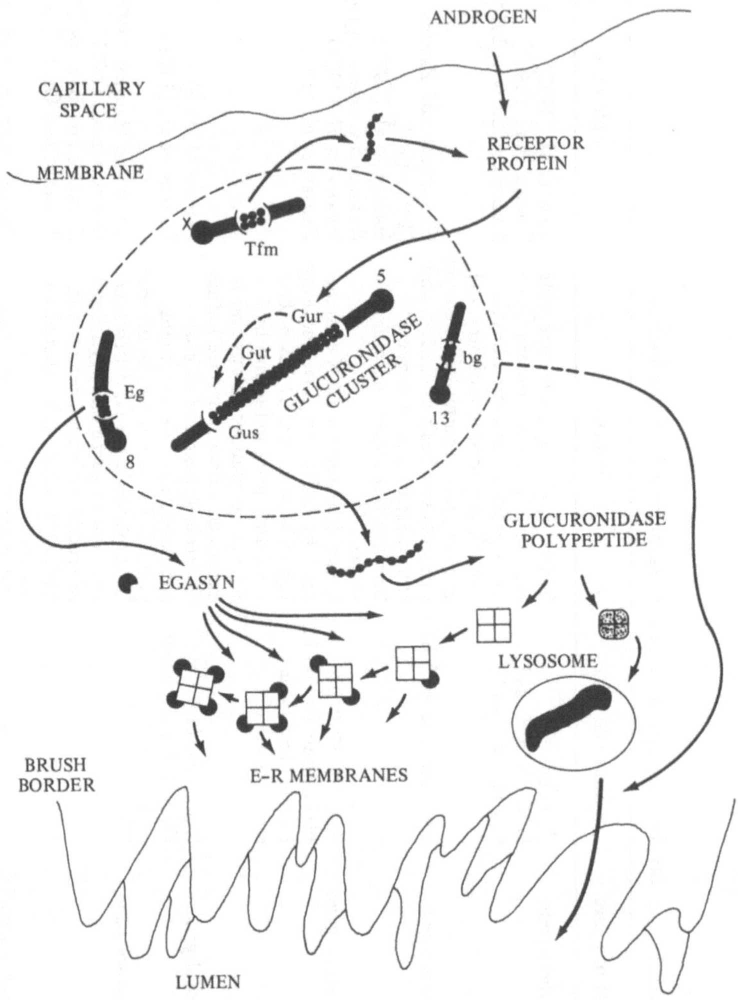

Fig. 10. Genetic control of β-glucuronidase in proximal tubule of mouse kidney. Figure is taken from PAIGEN et al. (1975) J. Cell. Physiol. **85**, 379

by certain stereoisomers of 5α-androstanediol; seemingly, this rapid induction does not require the receptor mechanism and is refractory to the antiandrogen, cyproterone acetate (OHNO and LYON, 1970; OHNO et al., 1971). By contrast, the induction of β-glucuronidase evoked by androgens in other reports (SWANK and PAIGEN, 1973; SWANK et al., 1973) took up to two weeks of androgen injections to promote the maximal response and clearly proceeds by genetic (or transcriptional) mechanisms.

In the present author's opinion, the superb concepts proposed by PAIGEN et al. (1975) highlight all the desirable features for future work on the mechanism of action of androgens. First, the use of mutants adds an exciting dimension to this work, permitting the identification of the specific genes involved in the androgenic regulation of a well-characterised macromolecule; this laudable success is unique among androgen target cells. The *gur* locus is solely implicated in the control of enzyme induction in mouse kidney and represents the first genetic element, sensitive to androgenic stimulation, to be identified unequivocally. Second, the work on kidney glucuronidase provides a novel insight into the developmental aspects of hormonal responses and as emphasized throughout this monograph, this is a particularly important aspect of the mechanism of action of androgens. Third, the work of PAIGEN et al. (1975) prompts a reappraisal of our thinking on the widely accepted similarities of the regulatory mechanisms in bacteria and mammalian cells. It is fashionable to consider that prokaryotic control mechanisms invariably have a similar if not identical counterpart in higher animals, but this may not be a valid assumption. Variants of the *gur* locus in mouse kidney are a good example in this context. The induction of β-glucuronidase may vary markedly in terms of rate and extent, but basal level of enzyme activity in the absence of androgens is the same in all recombinant mouse strains. This situation has no strict parallel in bacteria.

4. Morphologic Aspects of Androgenic Stimulation

This is a suitable topic for a monograph in its own right, but some discussion is warranted on two grounds. First, morphologic criteria substantiate the biochemical changes promoted by androgens because many of the inducible enzymes are associated with a particular intracellular organelle. Second, only morphologic studies provide the real extent of the involvement of androgens in maintaining the growth and function of many androgen target cells, particularly the male accessory sexual glands. Such trophic effects are not an invariant feature of the mechanism of action of androgens as, for example, in the anterior pituitary and hypothalamus. In addition, many of the biochemical responses evoked by androgens have no requirement for concomitant cell division or DNA replication, appropriate examples being the stimulation of nuclear protein synthesis (CHUNG and COFFEY, 1971 b) and the induction of kidney β-glucuronidase (OHNO et al., 1971). Also hypertrophy of mouse kidney can be stimulated by androgens without any appearance of hyperplasia (TETTENBORN et al., 1971).

A survey of the morphologic changes promoted by androgens in various target organs is presented in Table 15. The importance of the secretion of testosterone from the foetal testis was emphasized in Chapter II.3a and this is reflected in the morphologic changes occurring during the early developmental period. The classical observation on foetal morphology is credited to LILLIE (1917) who was the first to observe the masculinisation of the female in heterosexual twins owing to androgen secretion by the foetal testis of the male twin. The progeny are a normal male and a sterile freemartin. FLICKINGER (1974) has rightly stressed, however, that the changes in fine structure occurring in the foetal period are much less dramatic than in the postnatal period, particularly during the prodigious growth of the male accessory sexual glands during the attainment of sexual maturity. Certain changes in morphology occur during the neonatal period of development as a result of the secretion of androgens by the testis, including a modest increase in the membranes of the endoplasmic reticulum and the appearance of numerous and extensive pores in the nuclear membrane (DEANE and WURTZELMANN, 1965). The latter changes would facilitate nuclear-cytoplasmic interactions. In adult development, the morphologic changes promoted by androgens in the accessory sexual glands can only be described as overwhelming. The selected observations contained in Table 13 do not do justice to the wealth of literature available; for more details, the excellent surveys by BRANDES (1974 a, b) are strongly recommended. Many of the enzymes induced by androgens (see Table 12) have also been studied by sophisticated cytochemical procedures (BRANDES, 1974 a, b). These studies corroborate the extensive control of enzyme activities by androgens, as established by biochemical analysis. In contrast to the other male acces-

Table 15. Some of morphologic changes promoted by androgens in target cells.

Developmental stage	Organelle	Species	Effect	Reference
Foetal development	General	Cow	Sterile female (freemartin) in heterosexual twins.	LILLIE (1917)
	Sexual primordia	Rat	Ablation of foetal testis impairs development.	WELLS (1950)
	(prostatic cords, etc.)	Rabbit		JOST (1950)
		Rat	Androgens promote differentiation *in vitro*.	PRICE and PANNABECKER (1959)
	Early membranes	Rat	Anullate lamellae [a] dispersed by androgens.	van de HELDE and van de HELDE (1968)
	Endoplasmic reticulum	Rat	Stimulated by foetal androgens.	FLICKINGER (1970 a)
	Hemidesmosomes	Rat	Binds epithelium to connective tissue.	FLICKINGER (1970 b)
Neonatal development	Golgi apparatus	Rat	Enhanced together after birth and cisternae	DEANE and WURZELMANN (1965)
	Endoplasmic reticulum	Mouse	and secretory vacuoles become prominent	FLICKINGER (1971)
	Lysosomes	Rat	and larger in size.	FLICKINGER (1971)
Adult development	Lysosomes (kidney)	Mouse	Multiplication and lamellation in enzyme	PAIGEN et al. (1975)
	Lysosomes (others)		induction.	BRANDES (1974 a)
	Endoplasmic reticulum		Replacement by autophagic vacuoles after castration.	BRANDES and GROTH (1961, 1963) HELMINEN and ERICSSON (1970)
			Rough form, rich in ribosomes, requires the presence of androgens. The turnover correlates with protein synthesis; replaced by autophagic vacuoles after castration.	
	Golgi apparatus		Not active in castrated animals; necessary for the processing of secretory vacuoles (glycoproteins).	PETERSON and LEBLOND (1964) WHALEY et al. (1972)
		All species [b]		
	Mitochondria		Depleted, without cristae, in castrated animals.	ORLANDINI (1964; ALLISON (1964)
	Nucleus, nucleolus		Nuclear membrane distorts and active chromatin disappears after castration; chromatin in perinuclear area without androgens.	BRANDES (1974 a)
	Absorptive organelles (epididymis only)		Endocytotic vacuoles disappear after castration.	NICANDER (1965); FRIEND (1969)

[a] function of these membranes is not certain, but their synthesis proceeds androgen secretion; [b] male accessory sexual glands from all species respond similarly.

sory sexual glands, the epididymis has an absorptive as against a secretory function and this is reflected in its morphology (BRANDES, 1974 b); these absorptive processes are similarly maintained by high circulating concentrations of testosterone. The omnipotence of the trophic effects of androgens can also be demonstrated by the administration of the antiandrogen, cyproterone acetate, to pregnant animals (for deleterious foetal development; JOST, 1967) and to adult animals (DAHL and TVETER, 1974). A particularly precise approach to the morphologic changes evoked by androgens is to maintain the target cells as organ cultures and to supplement the chemically-defined medium with androgenic steroids (GITTINGER and LASNITSKI, 1972).

As an overview, the morphology of the microsomes, lysosomes, nuclei, mitochondria, and Golgi apparatus of the classical androgen target cells is maintained by androgens. For this reason, it is not surprising that the growth, structure, and function of the male accessory sexual glands mandatorily requires the continuous presence of circulating concentrations of testosterone in the adult.

Gross changes in the architecture of the male accessory sexual glands occur during the ageing process (MAINWARING and BRANDES, 1974), particularly in the deposition of the "ageing pigment," lipofuchsin. These lipofuchsin granules may be breakdown products of defective mitochondria (DUNCAN et al., 1960) but it is more likely that they are remnants of defunct Golgi elements (GATENBY, 1953; BONDAREFF, 1964). The morphologic changes in the ageing mouse prostate are correlated with an impairment of macromolecular synthesis in the nuclei (MAINWARING, 1968) and microsomes (MAINWARING, 1969 c). However, not all of the changes observed during senescence can be attributed to a limitation of the androgenic milieu because the administration of testosterone to old mice does not evoke profound biochemical changes (MAINWARING, 1967).

5. Conclusions

Early responses to androgens are responsible for the maintenance of the structure and function of androgen target cells, as typified by the male accessory sexual glands. The morphologic effects on androgens in the brain are much less pronounced and these differences should be explored rather than ignored in future work as they may give some clue to the molecular mechanisms involved in androgenic responses.

Macromolecular constituents of all the intracellular organelles are regulated by androgens and the induction of enzymes appears to proceed by amplification mechanisms. Androgens essentially regulate the synthesis of proteins by the provision of more messenger RNA and functional ribosomes; this implies that an enhancement of genetic transcription is the predominant feature of androgenic responses but translational control is also indicated in some studies (ICHII et al., 1974; MAINWARING et al., 1974 b).

In order to gain a further understanding into the regulation of macromolecular synthesis by androgens, it would seem that experimental systems other than rat ventral prostate gland will feature prominently in future work. In particular, the impressive advances made by PAIGEN et al. (1975) on the androgenic regulation of β-glucuronidase in mouse kidney emphasize the need for the wider employment of mutants in contemporary research on steroid hormones. Only by such means can future progress be ensured.

V. Late Events in the Mechanism of Action of Androgens

In this chapter, the regulation of DNA replication and cell division by androgens will be reviewed. Biochemical processes associated with cell renewal are evident only after a protracted period of androgenic stimulation and hence may aptly be described as the late androgenic responses. Androgens do not have a mitogenic action on all target cells, especially in the brain, and the present chapter is almost exclusively concerned with the male accessory sexual glands, particularly the rat ventral prostate gland. Together with the androgen receptor mechanisms described in Chapter III, studies on DNA replication are extremely important in the context of human prostatic hyperplasia and carcinoma. This work may lead to improvements in diagnosis and chemotherapy and hence will continue to feature prominently in future studies on the mechanism of action of androgens.

1. Cell Renewal and Morphologic Studies on the Mitogenic Activity of Testosterone and Its Metabolites

Since the secondary sexual glands in male animals were first recognised by their atrophy in castrated animals, their continued growth and maintenance has long been appreciated as a prominent feature of the biology of testosterone. Indeed, before biochemistry made such an impact on research on androgens, morphologic approaches were widely practiced. At first, the growth elicited by androgens was measured in rather loose, descriptive terms, either in the reduction in the size and weight of male accessory sexual glands in castrated animals or the reversal of these changes after the injection of androgenic steroids. From such studies, and especially the work of SAUNDERS (1963), a basic model for studying the involvement of androgens in mitosis was evolved. Some 7 or 10 days after bilateral orchidectomy, the male accessory sexual glands of rodents shrink to their minimum and the administration of testosterone then promotes a dramatic increase in their weight and intracellular secretions, but only after a protracted delay or latent period of 2 or 3 days. With rat ventral prostate as the model target organ, this system has been the basis for virtually all studies on DNA replication.

In early studies, the stimulation of mitosis in the prostate gland was monitored by histologic procedures (BURKHART, 1942; ALLEN, 1958), in which colchicine was injected, *in vivo,* to arrest dividing cells in metaphase and after histologic preparation of the specimens, the mitotic index was determined as the percentage of metaphase plates in the total number of cells observed. Studies such as these established one very important point, namely that the number of cells in the prostate glands of castrated animals was severely reduced and androgens were necessary for initiating cell renewal. The other observation was that mitotic figures or dividing cells were scarcely if ever seen in normal, noncastrated animals (mitotic index 0.001–0.01). However, mitotic indices give only a lower estimate of cell renewal and the more refined approach is to employ labelling indices where the injection of [^{3}H] thymidine is followed by autoradiography of tissue slices. This approach enables DNA replication and cell renewal to be measured together (TUOHIMAA and NIEMI, 1968; MORLEY et al., 1973). The labelling index is the percentage of cells observed to contain silver grains at any one time and in resting, nondividing glands it is higher than the mitotic index at a value of approximately 0.27–1.3 (TUOHIMAA and NIEMI, 1974). Using this autoradiographic approach, TUOHIMAA and NIEMI (1968) studied cell renewal in the prostate and seminal vesicle of castrated rats which had subsequently been given daily injections of testosterone. They demonstrated that mitotic activity rose to a peak 2 days after the beginning of hormonal stimulation, followed by a further but slight activation 2 days later. The two important features of this study were as follows. First, there is a clear lag in the process of cell renewal, the latent period lasting some 2 days. Second, in contrast to the response of female accessory sexual glands to oestrogenic stimulation (BULLOGH, 1965; BRESCIANI, 1971; LEE, 1971), there was no cyclical response in the mitotic activity in the prostate and seminal vesicle to testosterone. This simplifies the interpretation of experimental results very considerably.

TUOHIMAA and NIEMI (1968) also concluded that testosterone was capable of significantly shortening all phases of the cell cycle in androgen target cells. Based on the classical model of cell division proposed by HOWARD and PELC (1951), androgens particularly regulated the duration of the G1 phase. However, two of my colleagues (SMITH and MARTIN, 1973) have recently proposed a new concept for cell division in which cells may exist either in an indeterminate A state or a determinate B state. In this contemporary model, androgens would increase the probability of cells entering the B state and hence the process of mitosis. NIEMI and TUOHIMAA (1971) have established that although cell renewal in all male accessory sexual glands is stimulated by androgens, this proceeds at markedly different rates from one organ to another. Labelling indices were highest in seminal vesicle, in the middle range in prostate and lowest in certain segments of epididymis.

Despite the protracted latent period in the renewal of cells in rat prostate, the process is initially triggered by the androgen receptor mechanism and the high-affinity binding of 5α-dihydrotestosterone. This active metabolite seems to be a more potent mitogenic agent than testosterone, especially in the response of the prostate in organ culture to selected steroids *in vitro* (BAULIEU et al., 1968; GITTINGER and LASNITZKI, 1972; ROY et al., 1972 a, b). In addition, nuclear morphology is disorganised by the administration of cyproterone acetate to castrated rats, *in vivo* (DAHL and TVETER, 1974). For these reasons, the report by BRANDES (1974 b) that cyclic AMP can mimic the action of testosterone in the restitution of the structure of rat prostate and, by inference, cell renewal, is extremely surprising. This conclusion is diametrically opposed to the experimental findings from this laboratory (MANGAN et al., 1973) and elsewhere (CRAVEN et al., 1974). Processes mediated by cyclic AMP in the prostate gland are unique in that they are refractory to cyproterone acetate (MANGAN et al., 1973) and on the further evidence of CRAVEN et al. (1974) the second messenger concept is not applicable to prostate growth or cell renewal. To validate this point MANGAN et al. (1973) were unable to demonstrate any stimulation of prostatic growth or the restoration of the histologic features of the gland by the administration of cyclic AMP to castrated animals, *in vivo*. Our results are shown in Figure 11, and the disparity with the work of BRANDES (1974 b) remains baffling and totally unexplained.

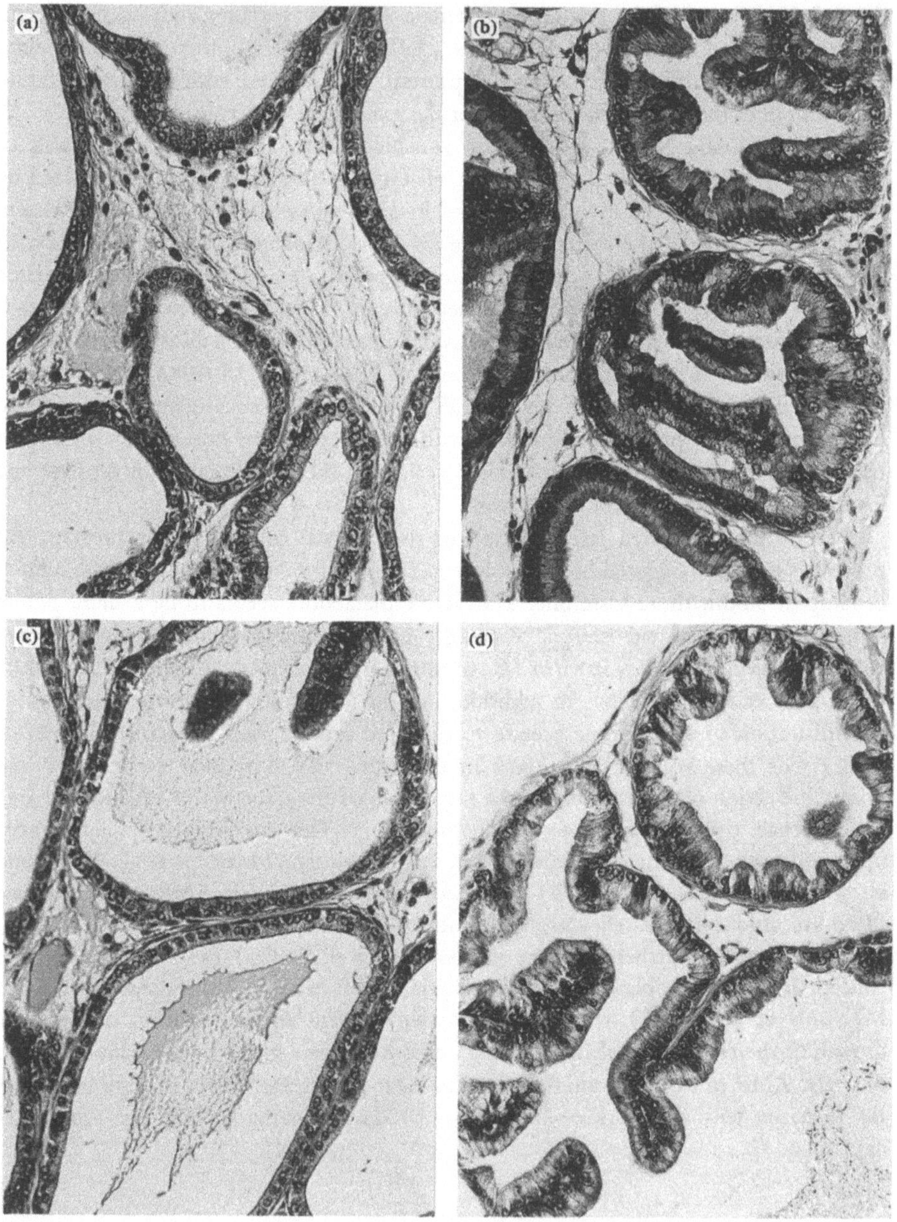

Fig. 11 a–d. Sections of prostate, stained with haemotoxylin and eosin. (a) From castrated animals, 72 h after orchidectomy; (b) from normal, noncastrated animals; (c) from castrated animals, after 1 day of treatment with 2.5 mg each of cyclic AMP and theophylline; (d) from castrated animals, after 1 day of treatment with 2.5 mg of testosterone phenylpropionate. Pictures are taken from MANGAN et al. (1973) Biochem. J. 134, 129

2. Biochemical Approaches to the Regulation of Mitosis in Rat Ventral Prostate Gland

Apart from the studies conducted several years ago on rat prostate (COFFEY et al., 1968) and rat coagulating gland (BRASEL et al., 1968) the biochemistry of DNA replication and mitosis in male accessory sexual glands has received little attention. Prompted, however, by the importance of such studies to human prostatic neoplasms, this laboratory has recently devoted much attention to the androgenic regulation of DNA replication. Based on the work of SAUNDERS (1963) and COFFEY et al. (1968), the experimental system has been the ventral prostate of rats castrated 7 days previously and subsequently maintained on daily injections of testosterone. Biochemical events associated with DNA replication and mitosis should demonstrate a latent period of 2 days despite the presence of androgens and then a dramatic surge of activity 3 or 4 days after the commencement of hormonal stimulation. This characteristic pattern of response makes the male accessory sexual glands ideal systems for investigations on DNA replication. Based mainly on the work of SUFRIN and COFFEY (1973), the binding of 5α-dihydrotestosterone must occur in the latent period for DNA synthesis to occur later.

a) Enzymes Engaged in DNA Replication

A diversity of enzymes are necessary for DNA replication in eukaryotes and their characterisation remains the centre of considerable debate (for a review, see KEIR and CRAIG, 1973). The DNA polymerases, catalysing the polymerisation of 2-deoxyribonucleoside 5′-triphosphates according to the nucleotide sequence of the DNA template, exist in several structurally distinct forms but mainly as a cytoplasmic 6 to 8S form, and a nuclear species, 3 to 4S (HAINES et al., 1971, 1972). Of these, only the 6 to 8S form is stimulated by the onset of cell division (CHANG and BOLLUM, 1972; CHANG et al., 1973) and this has been particularly well demonstrated in the division cycle of HeLa cells (CHIU and BARIL, 1975); the larger 9S form of polymerase was maximally active only during the S phase. An additional complication is that these forms of DNA polymerase may be interconvertible oligomers containing the same polypeptide subunit (LAZARUS and KITRON, 1973; HECHT, 1973) but this now seems less likely (CHIU and BARIL, 1975). Terminal nucleotidyltransferases promote the random addition of deoxyribonucleoside triphosphates to the 3′-terminus of DNA (CHANG and BOLLUM, 1971) but they are of extremely limited distribution (BEKKERING-KUYLAARS and CAMPAGNARI, 1974; PENI et al., 1974) and their involvement in DNA replication is far from proven (TATÒ et al., 1974). DNA ligases join short fragments of DNA together and these have been implicated in the mitotic process in mammalian cells (HAYTON et al., 1973). A little surprisingly, LEHMAN (1967) has postulated that deoxyribonucleases were needed for DNA replication in eukaryotes but this has never been rigorously confirmed. It has also long been recognised that thymidine kinase is a good indicator of the onset of mitosis (BOLLUM and POTTER, 1959; MCAUSLAN and JOKLIK, 1962) and this is now explained by the fact that intracellular pools of deoxyribonucleoside phosphates, including thymidine monophosphate, fluctuate enormously during the cell cycle, being maximal during the S phase (NORDENSKJÖLD et al., 1970).

On this evidence, the identification of the enzymes mandatorily engaged in DNA replication in rat prostate is a daunting task, particularly since so many of the possible enzymes exist in oligomeric forms. However, using methods of high resolution RENNIE et al. (1975) succeeded in unequivocally identifying many of the enzymes involved in DNA replication and a summary of this work is presented in Table 16. The most dramatic correlation of an enzyme activity with the onset of DNA replication was seen in a 9S form of DNA polymerase which showed a marked preference for single-stranded (denatured) DNA as template. This observation instigated considerable research on the strand separation of native DNA in the intact cell; this is discussed at length in section V.2.c. Additional studies by RENNIE et al. (1975) showed that the enhancement of all these enzyme activities was initiated by the androgen receptor system. First, the responses were acutely tissue-specific and could not be demonstrated in liver and spleen. Second, the responses were exceedingly steroid-specific and reflected precisely the selective binding of steroids to the receptor system. In particular, the enzyme activities were not enhanced directly by the administration of either cyproterone acetate or corticosterone, but an important distinction became evident when these two steroids were injected concomitantly with testosterone; only the antiandrogen suppressed the stimulatory effect of testosterone. This observation is in harmony with the work of SUFRIN and COFFEY (1973) who have suggested that the relative efficacy of antiandrogens may be judged by their suppression of the induction of DNA polymerase activity by testosterone under conditions *in vivo*. Third, the injection of actinomycin D or cycloheximide with testosterone totally suppressed the stimulation of all the enzymes associated with DNA replication, suggesting that continued RNA and protein synthesis was necessary for these late androgenic responses. By contrast, CHUNG and COFFEY (1971 b) suggest that RNA synthesis *de novo* is not an obligatory requirement for the stimulation of DNA polymerase activity.

In marked contrast to other enzymes induced by androgens and classified as early rather than late responses, as here, the molecular basis for the dramatic appearance of the enzymes associated with mitosis only after the long latent period is far from understood. It remains possible that the structural genes associated with DNA replication are closely grouped as a functional unit or genetic cluster, but that they are inaccessible or blocked by protein and available for transcription only when the DNA is fully exposed by the opening of the chromosomes just prior to DNA replication and mitosis. There are precedents for such events in embryonic development in amphibia (GURDON and WOODLAND, 1968), in the terminal stages of erythrocyte differentiation (DENTON et al., 1975), in the appearance of δ-aminolevulinate synthetase in the developing chick blastoderm (IRVING et al., 1976) and in the induc-

Table 16. Enzymes engaged in DNA replication in rat ventral prostate gland.

Numbers of enzymes are those recommended by International Union of Biochemistry (1972). Results are taken from RENNIE et al. (1975)

Enzyme	Method of analysis	Remarks
DNA polymerase (2.7.7.7)	Sucrose density gradients	Only 9S form, with marked preference for denatured [a] DNA as template, is involved in DNA replication. A 4S form, equally active with native or denatured DNA, was not influenced by androgens
Terminal transferases (2.7.7.7)	Sucrose density gradients	Not involved, since unaffected by castration.
Thymidine kinase (2.7.1.75)	Polyacrylamide gel electrophoresis [b]	Required for replication and two forms present; one form appears only after androgenic stimulation.
Deoxyribonucleases [c] (3.1.4.5)	Isoelectric focusing in polyacrylamide gels [d]	Many forms of enzyme were present but only one form (pI 7.0) associated with DNA replication.
DNA ligases (6.5.1.1)	Sucrose density gradients	Difficult to assay in crude extracts, but 10S form was associated with mitosis.

[a] denatured by heating at 100° C for 10 min, followed by rapid cooling to 0° C; [b] procedure of KIT et al. (1974) was used in presence of ATP to stabilize enzymes; [c] both types I and II were studied, maximally active at pH 8.0 (with 5 mM $MgCl^2$) and pH 5.0 (without divalent cations), respectively; [d] method was based on ZÖLLNER et al. (1974) where $[^3H]$-DNA is trapped in gel and enzyme activity is located by release of radioactivity into soluble form.

tion of tyrosine aminotransferase in HTC cells by glucocorticoids (MARTIN et al., 1969). In all instances, the appearance of certain proteins coincided precisely with an intense period of mitotic activity and gross morphologic change. This important aspect of the genetic regulation of enzymes associated with DNA replication warrants close attention in the future.

b) Proteins Regulating the Onset of Mitosis

As was first demonstrated in bacteria, many regulatory proteins have a characteristic ability to bind strongly to DNA (for a review, see VON HIPPEL and MCGHEE, 1972). Such a propensity for DNA is also believed to be a feature of the proteins engaged in maintaining the structure of the chromosomes and in regulating genetic expression in eukaryotic cells (KLEINSMITH et al., 1970; TENG et al., 1971; KLEINSMITH, 1973). As an extension of this early work, the analysis of proteins with a high affinity for DNA (termed, simply, DNA-binding proteins) has been conducted at various stages of the cell cycle and at different rates of cellular growth in the hope of identifying the proteins that control the onset of DNA replication and mitosis. Good examples among eukaryotic cells include Chinese hamster CHO cells (FOX and PARDEE, 1971), mouse 3T6 fibroblasts (SALAS and GREEN, 1971), a heterotrophic marine alga, *Cryptothecodinium cohnii* (FRANKER et al., 1973), human AGMK cells infected with adenovirus type 2 where some of the proteins are virus-specific (VAN DER VLIET and LEVINE, 1973) and Novikoff hepatoma cells (JOHNSON et al., 1974). The best demonstration of the involvement of a specific DNA-binding protein in cell division is in temperature-sensitive mutants of the initiation sites of DNA replication in *Escherichia coli* (GUDIS et al., 1975).

Against this background MAINWARING et al. (1976 a) investigated the synthesis of the DNA-binding proteins in rat ventral prostate at times before and during DNA replication. Castrated rats were injected with testosterone, *in vivo,* and after various periods of androgenic stimulation, minces of prostate tissue were incubated with [^{35}S]-methionine, *in vitro,* under carefully selected conditions where the incorporation of labelled amino acids into protein was strictly proportional to the amount of tissue but to no other factor. The [^{35}S]-labelled proteins were fractionated on DNA-cellulose columns to isolate the DNA-binding proteins and these were further resolved by electrophoresis in thin slabs of polyacrylamide in the presence of sodium dodecyl sulphate. The protein bands were located by the dye, Coomassie Blue, and the stained gels were photographed. The gels were next dried *in vacuo* and autoradiography was carried out by contact with x-ray film. The photographic negatives and the developed autoradiographs were finally scanned in a microdensitometer to estimate quantitatively the amounts of proteins present and their degree of labelling with [^{35}S] methionine. An example of such an experiment is presented in Figure 12. On the evidence obtained by MAINWARING et al. (1976 a), the DNA-binding proteins of rat prostate have some very interesting properties indeed. (1) They are a complex group of proteins, whose individual components are distinguished only by precise methods of fractionation. (2) The DNA-binding proteins in the rat prostate are in a dynamic state of turnover and extremely sensitive to androgenic stimulation. Selected members of this group of proteins were synthesized only during the latent period prior to DNA

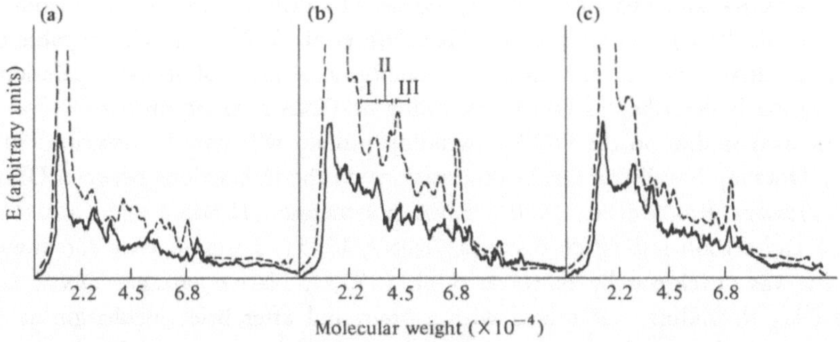

Fig. 12 a–c. Analysis of prostate DNA-binding proteins after various periods of androgenic stimulation. Beginning 7 days after castration, groups of 8 rats were injected daily with testosterone phenylpropionate (2.5 mg per animal). Pooled prostate glands were incubated with [^{35}S] methionine in vitro and labelled proteins with high affinity for DNA-cellulose were isolated. Samples of labelled protein (200 μg; 4000 to 1000 d.p.m. of [^{35}S]) were separated in thin slabs of polyacrylamide gel in presence of sodium dodecyl sulphate. Gels stained with Coomassie Blue were photographed, then dried *in vacuo* and autoradiography was carried out. The photographic negatives and autoradiographs were scanned in a microdensitometer. The bands I, II, and III were particularly influenced by androgens. Proteins from (a) castrated (control) animals; (b) after 24 h and (c) after 72 h of androgenic stimulation. Protein (stained gels)—; ^{35}S (autoradiographs)– –. Graphs are based on work of MAINWARING et al. (1976) Biochem. J. 156, 253

replication and mitosis, examples being the proteins designated I, II, and III in Figure 12. The changes in these DNA-binding proteins were tissue-specific and could not be detected in nontarget tissues, such as spleen and liver. (3) The synthesis of prostate DNA-binding proteins was prevented by the administration of cyproterone acetate *in vivo,* indicating that they are stringently regulated by the androgen receptor system. Taken overall, the DNA-binding proteins are likely candidates for inclusion among the intracellular regulators envisaged by FOX and PARDEE (1971), SALAS and GREEN (1971) and SUFRIN and COFFEY (1973) and which somehow control the entry of cells into mitosis.

c) DNA-Unwinding Proteins

The most striking feature of DNA replication in rat prostate was the dramatic appearance of a 9S form of DNA polymerase (see Table 16). However, since this type of enzyme is widely accepted as being responsible for DNA replication (CHANG and BOLLUM, 1972; CHANG et al., 1973; CHIU and BARIL, 1975) yet it has a uniquqe preference for single-stranded or denatured DNA as template, the critical question is how the prostate cell effects the necessary strand separation in the helical DNA template. Based on the original observations of WANG (1971) and the later studies by SIGAL et al. (1972) and MOLINEUX and GEFTER (1974), it is now evident that *E. coli* contains a protein capable of separating the strands of helical (or native) DNA. Until recently, studies on mammalian cells were singularly less advanced, but similar proteins have now been detected in human KB-3 cells (KELLER, 1975), calf thymus

(PULLEYBANK and MORGAN, 1975), mouse 3T3 cells infected with polyoma virus (YEH et al., 1976) and rat prostate (RENNIE et al., 1975). Proteins capable of unwinding native DNA or at least of introducing local areas of strand separation have been variously described as DNA-unwinding or DNA-relaxing proteins.

Our own studies on the DNA-unwinding protein will now be described in more detail. Drawing heavily on earlier observations on the interactions between DNA and *lac* repressor (RIGGS et al., 1970), RNA polymerase (HINKLE and CHAMBERLIN, 1972) and certain lysine-rich histones (RENZ, 1975), a simple assay for unwinding activity was developed by RENNIE et al. (1975). Native prostate DNA, labelled with [^{3}H] thymidine, was mixed with protein and after brief incubation at 37° C, the presence of unwinding activity was indicated by the retention of radioactivity on nitrocellulose (Millipore) membranes. Native, [^{3}H]-labelled DNA is not retained unless its secondary structure is modified by the introduction of local areas of strand separation. The following properties were ascribed to the prostate DNA-unwinding protein (RENNIE et al., 1975). (1) It is a representative of the prostate proteins that possess a high affinity for immobilised DNA and may therefore be readily isolated by DNA-cellulose chromatography. (2) The appearance of the DNA-unwinding protein was strictly regulated by androgens; its induction was selectively prevented by cyproterone acetate and a similar activity in spleen was insensitive to androgenic stimulation. (3) From time-course experiments, DNA-unwinding activity was maximally enhanced by androgens during the latent period prior to DNA replication.

Additional studies have since been carried out by MAINWARING et al. (1976a). A second method of assay has been developed which is more in keeping with the projected biological function of the DNA-unwinding protein of rat prostate. This improved assay is based on the preference of the 9S replicative form of DNA polymerase for denatured DNA. The enhancement of DNA synthesis with this particular polymerase on a template of native DNA gives a precise and sensitive assay of DNA-unwinding activity. To corroborate the specificity of the assay, it cannot be simulated with the 4S form of prostate DNA polymerase or the form I (Kornberg) enzyme from *E. coli,* both of which have a marked preference for native DNA as template, at least under the conditions of assay employed in our laboratory. Using this assay, all the distinctive features of the prostate DNA-unwinding activity were confirmed, particularly its acute regulation by androgens during the latent period prior to DNA replication. By a sequence of manipulations involving DNA-cellulose chromatography, DEAE-cellulose chromatography, sucrose gradient ultracentrifugation and isoelectric focusing in glycerol gradients, the prostate DNA-unwinding protein has been purified to homogeneity (650-fold purification) and its principal properties are summarised in Table 17. On present evidence, the prostate DNA-unwinding protein is a dimer composed of two identical subunits, molecular weight 40,000 each. It is also tempting to speculate that the protein corresponds to the DNA-binding protein designated as band II earlier in Figure 12.

Certain aspects of the prostate DNA-unwinding protein are worthy of particular comment. (1) In a manner analogous to its seeming counterpart in *E. coli* (MOLINEUX and GEFTER, 1974), the prostate DNA-unwinding forms a complex with the replicative 9S DNA polymerase that is sufficiently stable to be identified in sucrose gradients; it does show this behaviour toward the nonreplicative 4S DNA polymerase which has little preference for denatured DNA as template. (2) The change in the

Table 17. Properties of the DNA-unwinding protein of rat prostate gland.

DNA-unwinding activity may be monitored equally well by activation of native DNA for 9S replicative form of prostate DNA polymerase on retention of [^{3}H]-labelled native DNA on Millipore membranes. Results are taken from MAINWARING et al. (1976) Biochem. J. **156**, 263 and unpublished observations from this laboratory.

Property	Remarks
Hormonal specificity	Induced only by androgens; sensitive to antiandrogens
Distribution	Found in spleen, but here insensitive to androgens
Temporal appearance	In latent period, prior to DNA replication
Sedimentation coefficient	4.2S
Isoelectric point	8.4
Molecular weight	Approximately 80,000,[a] calculated from $S_{20,w}$ Approximately 40,000, in denaturing gels
Quaternary structure	Probably a dimer of 40,000 subunit
Interaction with native DNA	No change in sedimentation of prostate DNA Introduces areas in prostate DNA sensitive to S1 nuclease Relaxes supercoiled polyoma DNA
Interaction with DNA polymerase	Forms discrete 10.8S complex with 9S form of DNA polymerase; no interaction with 4S form
Interaction with native DNA and 9S DNA polymerase	Enables enzyme to form stable complex with native DNA

[a] calculated from equation of SCHACHMAN (1959), $S_A/S_B = (MW_A/MW_B)^{2/3}$

structure of native prostate DNA elicited by the unwinding protein must be relatively subtle, because no change in sedimentation behaviour of putatively "unwound" DNA was detectable. However, on incubation with [^{3}H]-labeled prostate DNA of high specific radioactivity, the unwinding protein promotes structural changes such that 10–15% of the radioactivity is sensitive to the S_1 nuclease of *Aspergillus oryzae;* this is a critical observation, because this enzyme is absolutely specific for single-stranded DNA (VOGT, 1973) and causes no detectable hydrolysis of native DNA in the absence of the unwinding protein. (3) Prostate DNA-unwinding protein also changes the structure of the supercoiled form of DNA from polyoma virus and the PM2 marine bacteriophage. These complex forms of DNA were first recognised by CHAMPOUX and DULBECCO (1972) and relaxation or unwinding of their structure may be detected by the binding of the fluorescent dye, ethidium bromide (PULLEYBANK and MORGAN, 1975), or a change in electrophoretic mobility in dilute agarose gels (SHARP et al., 1973). Like the other DNA-unwinding proteins derived from mammalian sources (KELLER, 1975; PULLEBANK and MORGAN, 1975), these sophisticated methods may be usefully applied to the prostate DNA-unwinding protein.

MAINWARING et al. (1976 a) conclude that the unwinding protein in rat prostate is induced at a time when native DNA must be converted into a single-stranded or relaxed form, capable of serving as the template for the replicative 9S form of DNA polymerase. Accordingly, this and other prostate proteins with a high affinity for

DNA (see section V.3.b) may regulate the entry of prostate cells into mitosis. Three outstanding aspects of the prostate DNA-unwinding protein must be tackled in the near future. First, the stoichiometry of its interaction with native DNA must be determined. Second, the relaxation of DNA should be monitored by electron microscopy. Third, the precise mechanism of strand separation in DNA should be ascertained.

d) Synthesis of Histones

This topic has not received much attention. WILLIAMS-ASHMAN and SHIMAZAKI (1967) reported that no significant change in the qualitative pattern of histones occurred during the stimulation of prostate growth by androgens. Later studies by CHUNG and COFFEY (1971 b) established that the synthesis of prostate histones, especially the lysine-rich histone F1, was controlled by androgens but to a lesser extent than other classes of nuclear proteins, notably those present in the nuclear membranes. Evidence from other systems overwhelmingly suggests that histone synthesis is intimately linked with DNA replication (ROBBINS and BORUN, 1967; GALLWITZ and MUELLER, 1969) and there is no a priori reason why prostate histones should be in any way different. The messenger RNA coding specifically for histones in eukaryotic cells is one of the extremely few that does not possess poly(A)-rich sequences at its 3′-terminus (ADESNIK and DARNELL, 1972; SCHOCHETMAN and PERRY, 1972). One can perceive no connection between this distinctive structural feature of histone messenger RNA and the fact that it is synthesized at the time of DNA replication. The real reason for the poly(A)-enriched region in most messenger RNA species is still not clear but it is certainly not a requisite for faithful translation of the message into protein (WILLIAMSON, et al., 1974). What is of interest is that the histone messenger RNA, like others coding for proteins with nuclear-directed functions, is transcribed only at the time of DNA replication.

3. Conclusions

Despite the long duration of the latent period in the response of many enzymes implicated in DNA replication to androgens, these responses are set in motion by the androgen receptor system. This conclusion is substantiated by the extreme specificity of these responses and their negation by antiandrogens. To a very large extent, the slow inception of the processes associated with DNA replication and mitosis may be explained by synthesis of the proteins which mandatorily regulate the onset of cell division. This concept has been propounded in the past (FOX and PARDEE, 1971; SALAS and GREEN, 1971) and the new evidence in the rat prostate is consistent with this viewpoint. In particular, a protein capable of activating native DNA by introducing local regions of strand separation is a possible representative of such regulatory proteins. The correlation between DNA replication and the appearance of many enzymes associated with this process raises the interesting possibility that a contiguous group of genetic loci may regulate these important nuclear events, including the synthesis of histones. It may be envisaged that this cluster of structural or other classification of genes is not available for transcription until the nuclear apparatus is dispersed or opened up at the time approaching mitosis. This possibility is currently under investigation in this laboratory.

Overall Conclusions

An attempt has been made in this monograph to provide a survey of contemporary progress and thinking on the mechanism of action of androgens. Unashamedly, this has been presented from a personal viewpoint but one that I hope the general reader will find both interesting and readable. Nevertheless, in a rapidly developing area of research such as this, individual prejudices often rise inadvertently to the surface, but the scope of the review has deliberately been kept as wide as possible.

The principal objective has been to describe all the events that follow the entry of testosterone into androgen target cells but in particular, to describe the biochemical responses that are set in motion by the advent of hormonal stimulation. For further details on the morphologic changes that follow androgenic stimulation, the reviews by BRANDES (1974 a, b) are particularly recommended. The deliberate emphasis in this review has been to describe the activation of cellular functions and biochemical processes once androgenic stimulation has been inititiated, rather than the alternative approach of describing the impairment of these parameters in castrated animals *in vivo* or androgen-depleted cells *in vitro*. The positive approach has two advantages. First, the androgenic responses are seen against the low functional state of the castrated animal with the additional benefit of the phenomenon of "biological overshoot," whereby a low level of response can be exaggerated once the appropriate stimulus is provided, followed by a decline to a resting level commensurate with the establishment of the steady state. As an example of such exploitation, it is unlikely that the prostate DNA-unwinding protein would have been discovered without the long latent period in DNA replication existing in castrated animals. Second, the difference in a given biochemical response is nearly always greater in the comparison of castrated animals with androgen-stimulated castrated animals rather than with normal (noncastrated) animals. Notwithstanding these considerations, studies on the involution of androgen target cells in castrated animals are very important, a good example being the work of BRUCHOVSKY and CRAVEN (1975) on the deleterious effects of hormonal deprivation on the androgen receptor mechanism.

The segregation of androgenic responses according to their relative rates of inception is by no means perfect, as it may be difficult to place a given activity into the "early" as against "late" category. Nevertheless, the temporal segregation of responses does provide a reasonable framework for discussion. What should be emphasized is that the initial and late responses can generally be elicited by a single injection of testosterone into castrated animals, whereas the late responses require repeated daily injections of androgens. Studies by ANDERSON et al. (1974) and especially by LAN and KATZENELLENBOGEN (1976) have thrown considerable light on the molecular basis for the temporal relationship between hormonal responses. The latter authors compared the effects of oestriol and its substituted forms, which resist steroid catabolism, on the early and late oestrogenic responses in the immature rat uterus. The phosphorylation of 2-deoxyglucose and DNA synthesis were taken as examples of the early and late responses (LAN and KATZENELLENBOGEN, 1976). The early response could be evoked by oestriol even though it saturated the nuclear binding sites for only a brief period of time, but importantly, the late response was not observed. On the other hand, oestriol cyclopentyl ether evoked both the early and late responses and was retained in the nucleus for the duration of the experiments. LAN and KATZENELLENBOGEN (1976) concluded that the sustained output theory rather than the alternative cascade or domino principle provides the best explanation of the hormonal dependence of late responses. The sustained output theory essentially depends on the persistent saturation of the nuclear binding mechanisms throughout late responses whereas the cascade principle suggests that once critical early events are completed, the late responses follow inexorably and even in the absence of the hormonal stimulus. These important findings are equally applicable to the late responses in the mechanism of action of androgens. Taking DNA replication in rat prostate as an appropriate example, this response requires persistent androgenic stimulation, it is sensitive to antiandrogens and it requires the stimulation of protein synthesis during the latent period (SUFRIN and COFFEY, 1972).

There are many indications that the halcyon days of studies on the mechanism of action of androgens are drawing to a close. This is not a statement of pessimism, but a prediction that future research objectives will be hard to achieve and will demand a more sophisticated approach than has been adopted hitherto. Studies on the metabolism of androgens are a good case in point. Such work has been universally popular for many years, largely because the necessary technology is highly advanced and the experiments are not demanding, either in terms of financial expenditure or personal effort. Metabolism will undoubtedly feature in future research but if the importance, specificity and function of androgens is to be elucidated in the brain, for example, then androgen metabolism must be studied in specified anatomical areas of the central nervous system. This will not be an easy undertaking.

To illustrate further the impending changes in the orientation of research, the rat ventral prostate gland will probably not continue in its present place of eminence because its limitations as a model system are now becoming apparent. If we are to ascertain the precise metabolic role of androgen receptor complexes, it is imperative that they are purified to a state approaching homogeneity. The small size of the rat prostate makes it unsuitable for such work and practical alternatives include the abundant nodules of human benign prostatic hyperplasia, the preen gland of the drake (WILSON and LOEB, 1965) and the submaxillary gland of both sexes of the pig (BOOTH, 1972). In addition, the androgenic responses in the rat prostate are extensive but insufficiently distinctive to provide a suitable platform for studies on the tissue specificity and genetic regulation of androgen-mediated phenomena. Nevertheless, the rat prostate may yet be the experimental system of choice in the solution of two outstanding aspects of the interactions of steroid hormones with target cells, namely the characterisation of the entry and exit mechanisms for androgens. The elucidation of the entry process is extremely important as it may provide new approaches to the chemotherapy of androgen-responsive neoplasms, but even in this instance, surgical samples of human prostatic carcinomata are more acceptable experimentally (GIORGI, 1976).

The most serious limitation to studies on the binding of androgens is that current methods are not sufficiently specific to identify unequivocally the restricted number of biologically important binding (or acceptor) sites in chromatin (YAMAMOTO and ALBERTS, 1975). This stricture is made even more onerous by our almost complete dependence on radioactive steroids as the means for identifying and measuring high-affinity binding sites. More effort must be directed toward the development of improved methods for measuring receptor sites and the wider use of [^{3}H] methyltrienolone (BONNE and RAYNAUD, 1975) will dramatically improve the accuracy of binding studies. More consideration will also have to be given to the choice of androgen target cells for binding studies. For example, far more would be gained in comparing the binding of androgens in mouse strains of high and low inducibility with respect to the androgen-sensitive kidney enzyme, β-glucuronidase, than merely repeating previous studies on the accessory sexual glands of exotic species. Investigators are now very conscious of the importance of the metabolism of androgens and this is particularly necessary where developmental aspects of androgen action are the subject of investigation.

Until reliable methods are available for the quantification of small but crucial numbers of acceptor sites in chromatin, recourse can be made to other experimental approaches. For example, the technology for the isolation, characterisation and assay of eukaryotic messenger RNA is now advanced and this may provide some insight into the maintenance of the tissue-specificity of androgenic responses. Of current models, the induction of the basic proteins in rat seminal vesicle by androgens (HIGGINS et al., 1976) has much to offer in this context. The particularly elegant studies of PAIGEN et al. (1975) illustrate the future importance of genetic variants in the elucidation of the mechanism of action of androgens. Provided that a sufficiently wide spectrum of variants is available, then the genetic elements engaged in the regulation of androgenic responses may be unequivocally established. An additional advantage of the genetic approach, as shown by PAIGEN et al. (1975), is that a penetrating appraisal of tissue specificity and other aspects of androgenic regulation may be made without the need to isolate and characterise specific species of messenger RNA. However, a necessary corollary to this premise is that the androgen-induced constituent must be a well-characterised macromolecule and preferably a polypeptide. Clinical aberrations of male sexual differentiation also provide ideal systems for studying the genetic regulation of androgenic responses. As reviewed by GOLDSTEIN and WILSON (1975), at least 18 genes are necessary for the prenatal development of the male phenotype and specific mutations may provide vital clues to the molecular basis for many clinical abnormalities. Human tissues will certainly feature prominently in future work because human fibroblasts may now be cultured successfully with full preservation of their androgen receptor mechanisms (KEENAN et al., 1975) and their ability to metabolise androgens (WILSON, 1975).

The distinctive responses of a given target cell to different classes of steroid hormones has been a topic of widespread interest and until recently, it was difficult to explain this biological phenomenon. This is particularly important in the mechanism of action of androgens where a wide spectrum of testosterone metabolites is formed within the target cells. It is now becoming clear that steroid hormones regulate unique sites in the genome and that the synthesis of receptor proteins is also subject to genetic regulation. The availability of suitable mutants has largely been responsible for the development of these concepts. In the Tfm mouse, for example, the oestrogen receptor mechanism is fully preserved but the androgen receptor mechanism is totally deleted (BULLOCK and BARDIN, 1975 b), indicating that the synthesis of receptor proteins is regulated by unique structural genes. Analysis of two human syndromes substantiates the view that testosterone and its principal metabolite, 5α-dihydrotestosterone, can have discrete biological functions (GOLDSTEIN and WILSON, 1975). These closely related steroids promote the differentiation of distinct regions in the male urogenital tract. In the familial incomplete male pseudohermaphrodite type 2 syndrome, a deficiency in the production of 5α-dihydrotestosterone impairs the development of the embryonic urogenital sinus and external genitalia; the Wolffian duct, however, develops normally because this process selectively requires testosterone. On the other hand, in the human testicular feminisation syndrome, the development of the entire male phenotype is suppressed because the genetic lesion negates a process common to the development of all regions of the urogenital tract. The work of RENNIE and BRUCHOVSKY (1972) and BRUCHOVSKY et al. (1975) has done a great deal to unravel the distinctive functions of testosterone and 5α-dihydrotestosterone in androgen target cells. It now seems probable that testosterone alone can occupy certain sites within chromatin whereas 5α-dihydrotestosterone associates with nuclear acceptor sites only in the obligatory presence of the androgen receptor protein. This distinction adds powerful weight to the argument that these two androgens have distinct biological functions even within the same cell, a concept originally proposed by BAULIEU et al. (1968) and confirmed in a sophisticated manner by GITTINGER and LASNITZKI (1972). Several lines of independent evidence support the contention that the specificity of cellular responses to various types of hormonal stimulus resides in the binding specificity of the cytoplasmic receptor proteins (DE KLOET et al., 1975; KONTULA et al., 1975; MAINWARING et al., 1976 b). In all cases, the failure of a steroid to bind effectively to the receptor system curtailed its biological activity. However, the explanation of cellular specificity solely in terms of receptor specificity is not totally sound. First, as discussed in detail in Chapter I.4b. many instances may be cited in the literature of redundant binding (BAXTER et al., 1971; O'FARELL and DANIEL, 1971; WILSON and GOLDSTEIN, 1972). In these cases, the high-affinity binding of a steroid hormone was not commensurate with a biological response. The nature of the nuclear events that are also involved in the inception of hormonal responses remains unclear but binding mechanisms alone cannot account for the cellular changes evoked by steroids. Second, the recent study by RUH and RUH (1975) suggests that the regulation of responses by the receptor systems may be more apparent than real. These workers examined the steroid specificity of the appareance of the uterine "induced protein," originally described by NOTIDES and GORSKI (1966) and widely believed to be stimulated exclusively by estrogens. However, RUH and RUH (1975) established that high concentrations of 5α-dihydrotestosterone could also enhance the induction of this uterine protein and this capability was sensitive to antioestrogens but not antiandrogens. The effect of 5α-dihydrotestosterone was specific as it could not be simulated by its 5β-stereoisomer or testosterone, except at much higher concentrations. Although 5α-dihydrotestosterone was a much less effective ligand than oestradiol-17β for the receptor system and was an inducer only at very high concentrations, this interesting study suggests that the specificity of the cytoplasmic receptors may not be as exclusive as was originally thought.

Apart from the fact that androgens do not invariably have a mitogenic function in all androgen target cells, the general mechanism of action of androgens proposed in Chapter I (Fig. 1) is widely applicable. The exceptions include the effects of androgens on liver and skeletal muscle. In the experience of most investigators (for example, ANDERSON and LIAO, 1968; MAINWARING, 1969 b), rat liver is a suitable control for the high-affinity binding of androgens in the male accessory sexual glands and under conditions *in vivo* or *in vitro,* liver has a questionable ability to retain androgens selectively, especially in the nucleus. The claim of an androgen "receptor" by MILIN and ROY (1973) is somewhat tenuous, as it was based on marginal differences in the binding of androgens in the livers of normal and pseudohermaphrodite rats. However, their putative receptor was not well characterised and may be equated with the aberrant and nonfunctional binding of androgens reported by WILSON and GOLDSTEIN (1972) in the submandibular gland of animals with a repressed male phenotype. A quite revolutionary insight into the regulatory function of androgens on rat liver is provided by the work of GUSTAFSSON et al. (1975). These investigators have provided very convincing evidence indeed that a specific, high-affinity binding mechanism for androstenedione (androst-4-en-3, 17-dione) is present in liver. This steroid has minimal androgenic activity and is generally considered to have little biological importance except as a prehormone for testosterone (HORTON and TAIT, 1966). Nevertheless, further studies by Gustafsson and his colleagues are awaited with considerable interest as they could open up new concepts in the importance of androgen metabolism, especially with respect to the neonatal imprinting of hepatic enzymes. The status of the binding of androgens in skeletal muscle still remains the centre of considerable debate. While most investigators have failed to demonstrate significant binding of androgens in muscle (for example, MAINWARING and MANGAN, 1973; KREIG et al., 1974), the binding of testosterone has been reported by JUNG and BAULIEU (1972) and MICHEL and BAULIEU (1974). Certainly, the binding of testosterone rather than 5α-dihydrotestosterone is consistent with the work of POWERS and FLORINI (1975) on the synthesis of DNA in muscle cells maintained *in vitro,* but most of the other evidence disfavours the binding of testosterone in the anabolic function of androgens. First, many workers have failed to demonstrate androgen binding in muscle and this cannot be easily dismissed as interlaboratory variations in methodology. Second, the extreme differences in the structures of potent anabolic steroids (norbolethone and stanazol) as compared with potent androgenic steroids (testosterone, dimethylnortestosterone) discounts a common mechanism if current theories of induced-fit and the geometric assembly of receptor sites are tenable. Third, responses in ribosomal RNA synthesis can be evoked in hours yet athletes must take anabolic steroids for weeks before their effects are evident. The work of STEINITZ et al. (1971) also suggests that anabolic and androgenic responses are mediated by different molecular mechanisms. Fourth, autoradiography has consistently failed to demonstrate the significant localisation of radioactivity in muscle after the administration of [^{3}H] testosterone, *in vivo* (TVETER and ATTRAMADAL, 1968, 1969; SAR et al., 1970) and certainly a nuclear retention of [^{3}H] androgens has yet to be reported. In the present author's view, the molecular basis for the anabolic function of testosterone remains in doubt, an opinion shared by KOCHAKIAN (1975).

The critical importance of nuclear or transcriptional events has become the dogma of the mechanism of action of steroid hormones, but this view should not be held as inviolate. The evidence for a certain measure of cytoplasmic or translational control was found in the induction of aldolase by androgens (MAINWARING et al., 1974 b) and the enhancement of foetal haemoglobin synthesis by aetiocholanolone and other 5β-reduced steroids (IRVING et al., 1976) cannot be explained simply on the grounds of the synthesis of globin messenger RNA. In the developing chick blastoderm, globin messenger RNA is freely available at all times yet is not translated even in the presence of steroid inducers. The importance of nuclear processes in the mediation of hormonal responses can be studied in target cells in culture by use of the drug cytocholasin B, which disorganises the microtubules and hence nuclear function (WESSELLS et al., 1971). Studies on anucleate cells have been widespread. Many cellular functions such as spreading, attachment, locomotion, pinocytosis and contact inhibition are maintained after treatment with cytocholasin B (GOLDMAN et al. 1973). On the other hand, the drug does not necessarily impair the nuclear binding of steroid hormones (GORSKI and RAKER, 1973), but it does suppress hormonal responses such as the stimulation of glucose metabolism in rat uterus by oestradiol-17β (GORSKI and RAKER, 1973) and the induction of tyrosine aminotransferase by dexamethasone in HTC cells (IVARIE et al., 1975). These studies emphasize the importance of nuclear mechanisms in hormonal responses but cytoplasmic mechanisms should not necessarily be discounted.

In molecular terms, the regulation of the synthesis of ovalbumin messenger RNA by oestrogens in chick oviduct is most impressive. Taking but two examples of the outstanding papers by O'Malley and his collaborators, oestrogens stimulate the number of initiation sites available in oviduct chromatin for the exogenous RNA polymerase of *E. coli* (SCHWARTZ et al., 1976) and a strict correlation exists between the occupation of chromatin acceptor sites by the oestrogen receptor complex and the number of ovalbumin messenger RNA molecules transcribed (KALIMI et al., 1976). In view of this laudable success, it is very tempting to apply the successful technology of O'Malley and his colleagues to other steroid-sensitive systems, including androgen target cells. However, certain considerations dictate a measure of caution. First, the induction of ovalbumin is a prime example of a switch mechanism and qualitative changes of this magnitude probably require profound alterations in the structure of oviduct chromatin. By contrast, with the possible exception of the foetal differentiation of the male urogenital tract, all androgenic responses involve amplification mechanisms where the change in genetic transcription may be more subtle. Second, on the evidence of REEDER (1973), MARYANKA and GOULD (1973), and STEGGLES et al. (1974), transcriptional changes in chromatin need not invariably be detected by bacterial RNA polymerase. Indeed, the activation of prostate chromatin by androgens may be detected only by eukaryotic RNA polymerase, present endogenously or supplied exogenously, but not by the highly purified bacterial enzyme (MAINWARING and JONES, 1975). Perhaps the real lesson is that each experimental system must be treated strictly on the basis of its individual merits.

The outstanding contribution by biological sciences over the last decade has been the elucidation of the machinery by which all forms of macromolecules are fabricated within the cell. How these complicated processes are regulated by the demands of development, growth and repair remains a fundamental problem of universal interest and importance. For these reasons, the future study of the mechanism of action of steroid hormones, including the androgens, is safely assured. However, the understandable and rightful demands for scientific relevance, added to economic pressures, will probably channel future efforts into areas of research with clear application to clinical problems. The quest for plausible explanations of the entry of steroids into cells, neonatal programming, and the regulation of DNA synthesis would satisfy these altruistic ideals. In terms of fundamental rather than applied contributions, my priorities would be the complete purification of androgen receptor complexes, coupled with studies on the important genetic regulators in androgen target cells.

When the Royal Society was formed in 1660, Robert Hooke, and other founder members pursued their science in an atmosphere of tranquility and gentlemanly reserve. Since then, the pace of science became a brisk walk, a jog, and now, a forced march. So much so, that as I wearily lay down my pen, many of the problems that I have outlined may even have been solved. So be it.

"When the dust has settled,
You will know whether you were riding a camel or an ass."
(Old Jewish proverb)

References

ADACHI, K., KANO, M.: The role of receptor proteins in controlling androgen action in the sebaceous glands of hamsters. Steroids **19**, 567 (1972).

ADESNIK, M., DARNELL, J. E.: Biogenesis and characterization of histone messenger RNA in HeLa cells. J. molec. Biol. **67**, 397 (1972).

AHMED, K.: Studies on nuclear phosphorproteins of rat ventral prostate: incorporation of ^{32}P from [γ-^{32}P]ATP. Biochim. biophys. Acta **243**, 38 (1971).

AHMED, K., WILLIAMS-ASHMAN, H. G.: Studies on the microsomal sodium-plus-potassium ion-stimulated adenosine triphosphatase in rat ventral prostate. Biochem. J. **113**, 829 (1969).

ALBERGA, A., FERREZ, M., BAULIEU, E-E.: Estradiol-receptor-DNA-interactions: liquid polymer phase separation. FEBS Lett. **61**, 223 (1976).

ALBERTS, B. M., FREY, L. M.: The bacteriophage gene 32: a structural protein in the replication and recombination of DNA. Nature (Lond.) **227**, 1313 (1970).

ALBERTSSON, P. Å.: Partition studies on nucleic acids I. Influence of electrolytes, polymer concentrations and nucleic acid conformations on the partition in the dextran-polyethylene glycol system. Biochim. biophys. Acta **103**, 1 (1965).

ALFHEIM, I., FJELL, B.: 5α-Dihydrotestosterone and the synthesis of RNA in the rat ventral prostate *in vitro*. II. Studies on the qualitative effect. Acta endocr. (Kbh.) **73**, 189 (1973).

ALLEN, J. M.: The influence of hormones on cell division. II. Time response of seminal vesicle, coagulating gland and ventral prostate of castrated male mice to a single injection of testosterone propionate. Exp. Cell. Res. **14**, 142 (1958).

ALLFREY, V. G.: Functional and molecular aspects of DNA-associated proteins. Histones and nucleohistones, p. 241, D. M. P. Phillips (ed.) London: Plenum Press 1971.

ALLISON, V. F.: Ultrastructural changes in the seminal vesicle epithelium of the rat following castration and androgen stimulation. Anat. Rec. **148**, 254 (1964).

ANDERSON, J. N., PECK, E. J., CLAR, J. H.: Nuclear receptor-estrogen complex: *in vivo* and *in vitro* binding of estradiol and estriol as influenced by serum albumin. J. Steroid Biochem. **5**, 103 (1974).

ANDERSON, K. M., LIAO, S.: Selective retention of dihydrotestosterone by prostatic nuclei. Nature (Lond.) **219**, 277 (1968).

ANDRÉ, J., ROCHEFORT, H.: Specific effect of estrogens on an interaction between the uterine estradiol receptor and DNA. FEBS Lett. **29**, 135 (1973).

ARIAS, F., SWEET, F., WARREN, J. C.: Affinity labelling of steroid binding sites. Study of the active site of 20β-hydroxysteroid dehydrogenase with 6β- and 11α bromacetoxy-progesterone. J. biol. Chem. **248**, 5641 (1973).

ARIAS, I. M., DOYLE, M., SCHIMKE, R. T.: Studies on the synthesis and degradation of proteins of the endoplasmic reticulum of rat liver. J. biol. Chem. **244**, 3303 (1969).

ARMSTRONG, D. T., KING, E. R.: Uterine progesterone metabolism and progestational response: effects of estrogens and prolactin. Endocrinology **89**, 1919 (1971).

ARTMAN, M., ROTH, J. S.: Chromosomal RNA: an artefact of preparation? J. molec. Biol. **60**, 291 (1971).

ATGER, M.: Studies on the entry of estrogens into uterine cells. J. Steroid Biochem. **5**, 342 (1974).

ATTARDI, B., OHNO, S.: Cytosol androgen receptor from kidney of normal and testicular feminized (Tfm) mice. Cell **2**, 205 (1974).

AVIV, H., LEDER, P.: Purification of biologically active globin messenger RNA by chromatography on oligothymidylic acid cellulose. Proc. nat. Acad. Sci. (Wash.) **69**, 1408 (1972).

AXEL, R., MELCHIOR, W., SOLLNER-WEBB, B., FELSENFELD, G.: Specific sites of interaction between histones and DNA in chromatin. Proc. nat. Acad. Sci. (Wash.) **71**, 4101 (1974).

AXÉN, R., PORATH, J., ERNBACH, S.: Chemical coupling of peptides and protein to polysaccharides by means of cyanogen halides. Nature (Lond.) **214**, 1302 (1967).

BAELEN, H. VAN, BECK, M., DE MOOR, P.: A cortisol-induced charge difference in desialylated human transcortin by isoelectric focusing. J. biol. Chem. **247**, 2699 (1972).

BAELEN, H. VAN, HEYNS, W., DE MOOR, P.: Microheterogeneity of testosterone-binding globulin in human pregnancy serum demonstrated by isolectric focusing. Ann. Endocr. (Paris) **30**, 199 (1969).

BAELEN, H. VAN, BECK, HEYNS, BAIRD, D., HORTON, R., LONGCOPE, C., TAIT, J. R.: Steroid dynamics under steady-state conditions. Recent Progr. Horm. Res. **25**, 611 (1969).

BALDWIN, J. P., BOSELEY, P. G., BRADBURY, E. M., IBEL, K.: The subunit structure of the eukaryotic chromosome. Nature (Lond.) **253**, 245 (1975).

BALLARD, P., WILLIAMS-ASHMAN, H. G.: Fractionation of the bulk proteins of seminal vesicle secretions. Invest. Urol. **2**, 38 (1964).

BARDIN, C. W., ALLISON, J. E., STANLEY, A. J., GUMBECK, L. G.: Secretion of testosterone by the pseudohermaphrodite rat. Endocrinology **84**, 435 (1969).

BARDIN, C. W., BULLOCK, L. P., SCHNEIDER, G., ALLISON, J. E., STANLEY, A. J.: Pseudohermaphrodite rat: endorgan insensitivity to testosterone. Science **167**, 1136 (1970).

BARLEY, J., GINSBURG, M., GREENSTEIN, B. D., MCLUSKY, N. J., THOMAS, B. J.: A receptor mediating sexual differentiation. Nature (Lond.) **252**, 259 (1974).

BARNEA, A., WEINSTEIN, A., LINDNER, H. R.: Uptake of androgens by the brain of the neo-natal female rat. Brain Res. **46**, 391 (1972).

BARRACLOUGH, C. A.: Modifications in reproductive function after exposure to hormones during the prenatal and early postnatal period. Neuroendocrinology, vol. 2, p. 69, L. Martini and W. F. Ganong (eds.). New York: Academic Press, 1967.

BASHIRELAHI, N., CHADER, G. J., VILLEE, C. A.: Effects of dihydrotestosterone on the synthesis of nucleic acid and ATP in prostatic nuclei. Biochem. biophys. Res. Commun. **37**, 976 (1969).

BASHIRELAHI, N., VILLEE, C. A.: Synthesis of nucleic acids by isolated nuclei: effect of testosterone and dihydrotestosterone. Biochim. biophys. Acta **202**, 192 (1970).

BAULIEU, E-E., JUNG, I.: A prostatic cytosol receptor. Biochem. biophys. Res. Commun. **38**, 599 (1970).

BAULIEU, E-E., JUNG, I., BLONDEAU, J-P., ROBEL, P.: Androgen receptors in rat ventral prostate. Advanc. Biosci. **7**, 179 (1971).

BAULIEU, E-E., LASNITZKI, I., ROBEL, P.: Metabolism of testosterone and actions of metabolites on prostate glands grown in organ culture. Nature (Lond.) **219**, 1155 (1968).

BAXTER, J. D., HARRIS, A. W., TOMKINS, G. M., COHN, M.: Glucocorticoid receptors in lymphoma cells in culture: relationship to glucocorticoid killing activity. Science **171**, 189 (1971).

BAXTER, J. D., TOMKINS, G. M.: Specific cytoplasmic glucocorticoid hormone receptors in hepatoma tissue culture cells. Proc. nat. Acad. Sci. (Wash.) **68**, 932 (1971).

BEHNKE, J. N., DAGHER, S. M., MASSEY, T. M., DEAL, W. C.: Rapid, multisample isoelectric focusing in sucrose density gradients using conventional polyacrylamide electrophoresis equipment. Analyt. Biochem. **69**, 1 (1975).

BEKKERING-KUYLAARS, S. A. M., CAMPAGNARI, F.: Characterization and properties of a DNA polymerase partially purified from the nuclei of calf thymus cells. Biochim. biophys. Acta **349**, 277 (1974).

BELL, P. A., MUNCK, A.: Steroid binding properties and stabilization of cytoplasmic glucocorticoid receptors from rat thymus cells. Biochem. J. **136**, 97 (1973).

BLANDEAU, R. J.: On the factors involved in sperm transport through the cervix uteri of the albino rat. Amer. J. Anat. **77**, 253 (1945).

BLAQUIER, J. A.: Selective uptake and metabolism of androgens by rat epididymis. The presence of a cytoplasmic receptor. Biochim. biophys. Res. Commun. **45**, 1076 (1971).

BLAQUIER, J. A., BREGER, D., CAMEO, M. S., CALANDRA, R.: The activation of cultured epididymal tubules by androgens. J. Steroid Biochem. **6**, 573 (1975).

BLOCH, E.: Metabolism of 4-^{14}C-progesterone by human fetal testis and ovaries. Endocrinology **74**, 833 (1964).

BOESEL, R. W., SHAIN, S. A.: A rapid, specific protocol for determination of available androgen receptor sites in unfractionated rat ventral prostate cytosol preparations. Biochem. biophys. Res. Commun. **61**, 1004 (1974).

BONDAREFF, W.: Histophysiology of the ageing nervous system. A. Geront. Res. **1**, 1 (1964).

BONNE, C., RAYNAUD, J-P.: Mode of spironolactone anti-androgenic action: inhibition of androstanolone binding to rat prostate androgen receptor. Molec. cell. Endocr. **2**, 59 (1974).

BONNE, C., RAYNAUD, J-P.: Methyltrienolone, a specific ligand for cellular androgen receptors. Steroids **26**, 227 (1975).

BOOTH, W. D.: The occurrence of testosterone and 5α-dihydrotestosterone in the submaxillary salivary gland of the boar. J. Endocr. **55**, 119 (1972).

BORIS, A., SCOTT, J. W., DE MARTINO, L., COX, D. C.: Endocrine profile of a nonsteroidal antiandrogen N-(3,5-dimethyl-4-isooxazoylmethyl)phthalimide (DIMP). Acta endocr. (Kbh.) **72**, 604 (1973).

BRADBURY, E. M., CRANE-ROBINSON, C.: Physical and conformational studies of histones and nucleohistones. Histones and nucleohistones, p. 85, D. M. P. Phillips (ed.). London: Plenum Press 1971.

BRADBURY, E. M., INGLIS, R. J., MATHEWS, H. R.: Control of cell division by very lysine-rich histone (Fl) phosphroylation. Nature (Lond.) **247**, 257 (1975 a).

BRADBURY, E. M., INGLIS, R. J., MATHEWS, H. R., LANGAN, T. A.: Molecular basis of control of mitotic cell division in eukaryotes. Nature (Lond.) **249**, 533 (1975 b).

BRADBURY, E. M., INGLIS, R. J., MATHEWS, H. R., SARNER, N.: Phosphorylation of very lysine-rich histone in *Physarum polycephalum*. Correlation with chromosome condensation. Europ. J. Biochem. **33**, 131 (1973).

BRANDES, D.: Fine structure and cytochemistry of male accessory sex accessory organs. Male accessory sex organs, p. 18, D. Brandes (ed.). New York: Academic Press 1974 a.

BRANDES, D.: Hormonal regulation of fine structure. Male accessory sex organs, p. 184, D. Brandes (ed.). New York: Academic Press 1974 b.

BRANDES, D., GROTH, D. P.: The fine structure of the rat prostatic complex. Exp. Cell Res. **23**, 159 (1961).

BRANDES, D., GROTH, D. P.: Functional ultrastructure of rat prostatic epithelium. Nat. Cancer Inst. Monogr. **12**, 47 (1963).

BRASEL, J. A., COFFEY, D. S., WILLIAMS-ASHMAN, H. G.: Androgen-induced changes in the DNA polymerase activity of coagulating glands of castrated rats. Med. Exp. **18**, 321 (1968).

BRECHER, P. O., CHABAUD, J-P., COLUCCI, V., DE SOMBRE, E. R., FLESHER, J. W., GUPTA, G. N., HUGHES, A., HURST, D. J., IDEDA, M., JACOBSON, H. I., JENSEN, E. V., JUNGBLUT, P. W., KAWASHIMA, T., KYSER, K. A., NEUMANN, H-G., NUMATA, M., PUCA, G. A., SAHA, N., SMITH, S., SUZUKI, T.: Estrogen receptor studies at the University of Chicago. Advanc. Biosci. **7**, 75 (1971).

BRESCIANI, F.: A variant steroid control of cell proliferation in the mammary gland

and cancer. Basic actions of sex steroids on target organs, p. 130, P. O. Hubinont, C. Leroy and P. Galand (eds.). Basel: S. Karger 1971.

BRINKMANN, A. O., MULDER, E., LAMERS-STAHLHOFEN, G. J. M., VAN DER MOLEN, H. J.: An oestradiol receptor in rat testis interstitial tissue. FEBS Lett. **26**, 301 (1972).

BRONSON, F. H., WHITSETT, J. M., HAMILTON, T. H.: Responsiveness of accessory glands of adult mice to testosterone: priming with neonatal injections. Endocrinology **90**, 10 (1972).

BROWN-GRANT, K., MUNCK, A., NAFTOLIN, F., SHERWOOD, M. R.: The effects of the administration of testosterone propionate alone or with phenobarbitone and testosterone metabolites on neonatal female rats. Horm. Behav. **2**, 173 (1971).

BRUCHOVSKY, N.: The metabolism of testosterone and dihydrotestosterone in an androgen-dependent tumour. A possible correlation between dihydrotestosterone and tumour growth, *in vivo*. Biochem. J. **127**, 561 (1972).

BRUCHOVSKY, N., CRAVEN, S.: Prostatic involution: effect on androgen receptors and intracellular androgen transport. Biochem. biophys. Res. Commun. **62**, 837 (1975).

BRUCHOVSKY, N., RENNIE, P. S., LESSER, B., SUTHERLAND, D. J. A.: The influence of androgen receptor on the concentration of androgens in nuclei of hormone-responsive cells. J. Steroid Biochem. **6**, 551 (1975).

BRUCHOVSKY, N., WILSON, J. D.: The conversion of testosterone to 5α-androstan-17β-ol-3-one by rat prostate *in vivo* and *in vitro*. J. biol. Chem. **243**, 2012 (1968).

BUCOURT, R., NEDELEC, L., TORELLI, V., GASC, J. C., VIGNAU, M.: An improved total synthesis of 2-oxa-19-norsteroids using a new Reformatsky reaction. Excerpta Medica Int. Congr. Ser. **219**, 125 (1971).

BULLOCK, L. P., BARDIN, C. W.: Decreased dihydrotestosterone retention by preputial gland nuclei from the androgen-insensitive pseudohermaphrodite rat. J. clin. Endocr. **31**, 113 (1970).

BULLOCK, L. P., BARDIN, C. W.: Androgen receptors in testicular feminisation. J. clin. Endocr. **35**, 935 (1972).

BULLOCK, L. P., BARDIN, C. W.: *In vivo* androgen retention in mouse kidney. Steroids **20**, 107 (1975 a).

BULLOCK, L. P., BARDIN, C. W.: The presence of estrogen receptor in kidneys from normal and androgen-insensitive Tfm/y mice. Endocrinology **97**, 1106 (1975 b).

BULLOCK, L. P., BARDIN, C. W., GRAM, T. E., SCHROEDER, D. H., GILLETTE, J. R.: Hepatic ethyl morphine demethylase and Δ^4 steroid reductase in the androgen-insensitive pseudohermaphrodite rat. Endocrinology **88**, 1521 (1971 a).

BULLOCK, L. P., BARDIN, C. W., OHNO, S.: The androgen insensitive mouse: absence of intranuclear androgen retention in the kidney. Biochem. biophys. Res. Commun. **44**, 1537 (1971 b).

BULLOCK, L. P., MAINWARING, W. I. P., BARDIN, C. W.: The physicochemical properties of the cytoplasmic androgen receptor in kidneys of normal, carrier female (Tfm/+) and androgen-insensitive (Tfm/y) mice. Endocr. Res. Commun. **2**, 25 (1975).

BULLOUGH, W. S.: Mitotic and functional homeostasis: a speculative review. Cancer Res. **25**, 1683 (1965).

BURGOYNE, L. A., HEWISH, D. R., MOBBS, J.: Mammalian chromatin substructure: studies with the calcium-magnesium endonuclease and two-dimensional polyacrylamide gel electrophoresis. Biochem. J. **143**, 67 (1974).

BURIC, C., BECKER, H., PETERSON, C., VOIGT, K. D.: Metabolism and mode of action of androgens in target tissues of male rats. Acta endocr. (Kbh.) **69**, 153 (1972).

BURKHART, E. Z.: A study of the early effect of androgenic substances in the rat by the aid of colchicine. J. exp. Zool. **89**, 135 (1942).

BURSTEIN, S. H.: The removal of testosterone-binding globulin from plasma by affinity chromatography. Steroids **14**, 263 (1969).

BUTLER, W. W. S., SCHADE, W. L.: The effect of castration and androgenic replacement on the nucleic acid composition, metabolism and enzymatic capabilities of the rat ventral prostate. Endocrinology **63**, 271 (1958).

CAMPBELL, H. J.: Effects of neonatal injections of hormones on sexual behaviour and reproduction in the rabbit. J. Physiol. (Lond.) **181**, 568 (1965).

CAMPBELL, J. A., LYSTER, S. C., DUNCAN, G. W., BABCOCK, J. C.: 7α-Methyl-19-norsteroids: a new class of potent anabolic and androgenic hormones. Steroids **1**, 317 (1963).

CAMPBELL, P. N., LAWFORD, G. R.: The protein-synthesizing activity of the endoplasmic reticulum in liver. Structure and function of the endoplasmic reticulum, p. 57, F. C. Gran, Ed. New York, Academic Press 1967.

CARTER, M. F., CHUNG, L. W. K., COFFEY, D. S.: A model system for screening anti-androgens. Urological Research, p. 27, L. R. King and G. P. Murphy (eds.), New York: Plenum Press 1972.

CASTEÑADA, E., LIAO, S.: The use of anti-steroid antibodies on the characterization of steroid receptors. J. biol. Chem. 250, 883 (1975).

CATT, K. J., DUFAU, M. L., NEAVES, W. B., WALSH, P. C., WILSON, J. D.: LH-hCG receptors and testosterone content during differentiation of the testis in the rabbit embryo. Endocrinology **97**, 1157 (1975).

CHAMBERLAIN, J., JAGARINEC, C., OFNER, P.: Catabolism of [4-^{14}C] testosterone by subcellular fractions of human prostate. Biochem. J. **99**, 610 (1966).

CHAMNESS, G. C., HENNINGS, A. W., MCGUIRE, W. L.: Oestrogen receptor binding is not restricted to target nuclei. Nature (Lond.) **241**, 458 (1973).

CHAMPOUX, J. J., DULBECCO, R.: An activity from mammalian cells that untwists super-helical DNA—a possible swivel for DNA replication. Proc. nat. Acad. Sci. (Wash.) **69**, 143 (1972).

CHANDLER, J. A., HARPER, M. E., BLUMDELL, G. K., MORTON, M. S.: Examination of the subcellular distribution of zinc in rat prostate after castration using the electron microscope microanalyser EMMA. J. Endocr. **65**, 34P (1975).

CHANG, L. M. S., BOLLUM, F. J.: Deoxynucleotide-polymerizing enzymes of the calf thymus gland. V. Homogeneous terminal deoxynucleotidyl-transferase. J. biol. Chem. **246**, 909 (1971).

CHANG, L. M. S., BOLLUM, F. J.: Variations of DNA polymerase activity during rat liver regeneration. J. biol. Chem. **247**, 7948 (1972).

CHANG, L. M. S., BROWN, M., BOLLUM, F. J.: Induction of DNA polymerase in mouse L cells. J. molec. Biol. **74**, 1 (1973).

CHAPDELAINE, A., DUPUY, G. M., ARTH, G. E., ROBERTS, K. D.: Inhibition of prostatic 5α-reductase and 3α-hydroxysteroid dehydrogenase activities. J. Steroid Biochem. **5**, 353 (1974).

CHIU, C. C., WARREN, J. C.: Synthesis of 2-diazoestrone sulphate and use for affinity labelling of steroid binding sites. Biochemistry **9**, 1917 (1970).

CHIU, R. W., BARIL, E. F.: Nuclear DNA polymerases and the HeLa cell cycle. J. biol. Chem. **250**, 7951 (1975).

CHUNG, L. W. K., COFFEY, D. S.: Biochemical characterization of prostatic nuclei. I. Androgen-induced changes in nuclear proteins. Biochim. biophys. Acta **247**, 570 (1971 a).

CHUNG, L. W. K., COFFEY, D. S.: Biochemical characterization of prostatic nuclei. II. Relationship between DNA synthesis and protein synthesis. Biochim. biophys. Acta **247**, 584 (1971 b).

CHUNG, L. W. K., FERLAND-RAYMOND, G.: Differences among rat sex accessory gland in their neonatal androgen dependency. Endocrinology **97**, 145 (1975).

CLARK, A. J.: The mode of action of drugs on cells. London: Arnold Press 1933.

CLEMENS, L. E., KLEINSMITH, L. J.: Specific binding of the oestradiol-receptor complex to DNA. Nature (Lond.) [New Biol.] 237, 204 (1972).

COFFEY, D. S., SHIMAZAKI, J., WILLIAMS-ASHMAN, H. G.: Polymerization of deoxy-ribonucleotides in relation to androgen-induced prostatic growth. Arch. Biochem. **124**, 184 (1968).

CONCHIE, J., FINDLAY, J.: Influence of gonadectomy, sex hormones and other factors on the activity of certain glycosidases in the rat and mouse. J. Endocr. **18**, 132 (1959).

COO-ER, A., LU, C. T., NORTEN, D. A.: The crystal and molecular structure of 9α-bromo-17β-hydroxy-17α-methylandrost-4-ene-3,11-dione. J. chem. Soc. Ser. B. 1228 (1968).

CORVOL, P., CHRAMBACH, A., RODBARD, D., BARDIN, C. W.: Physical properties and binding capacity of testosterone-oestradiol binding globulin in human plasma, determined by polyacrylamide gel electrophoresis. J. biol. Chem. 246, 3435 (1971).

CORVOL, P., MICHAUD, A., MENARD, J., FREIFELD, M., MAHOUDEAN, J.: Antiandrogenic effect of spirolactones: mechanism of action. Endocrinology 97, 52 (1975).

CRAVEN, S., LESSER, B., BRUCHOVSKY, N.: Evidence that adenosine 3',5'-cyclic monophosphate is not involved in the growth responses of the prostate to androgens. Endocrinology 97, 1177 (1974).

CUNHA, G. C.: Age-dependent loss of sensitivity of female urogenital sinus to androgenic stimulations as a function of epithelial-stromal interactions in mice. Endocrinology 95, 665 (1975).

DAHL, E., TVETER, K. J.: The ultrastructure of the accessory sexual organs of the male rat: the effect of cyproterone acetate. J. Endocr. 62, 251 (1974).

DALLNER, G., SIEKEVITZ, P., PALADE, G.: Biogenesis of endoplasmic reticulum membranes. 1. Structure and chemical differentiation. J. Cell Biol. 30, 73 (1966).

DANZO, B. J., ORGEBIN-CRIST, M-C., ELLER, B. C.: Changes in 5α-dihydrotestosterone binding to epididymal cytosol during sexual maturation in rabbits: correlation with morphological changes in testis and epididymis. Molec. cell. Endocr. 3, 203 (1975).

DANZO, B. J., ORGEBIN-CRIST, M-C., TOFT, D. O.: Characterization of a cytoplasmic receptor for 5α-dihydrotestosterone in the caput epididymis of the intact rabbit. Endocrinology 92, 310 (1973).

DARNELL, J. E., JELINEK, W. R., MOLLOY, G. R.: Biogenesis of messenger RNA: genetic regulation in mammalian cells. Science 181, 1215 (1973).

DAVID, K., DINGEMANSE, E., FREUD, J., LAQUEUR, E.: Über krystallinisches männliches Hormon aus Hoden (Testosteron) wirksamer auf aus Harn oder aus Cholesterin bereitetes Androsteron. Hoppe-Seylers Z. physiol. Chem. 233, 281 (1935).

DAVIES, P., FAHMY, A. R., PIERREPOINT, C. G., GRIFFITHS, K.: Hormone effects *in vitro* on prostatic ribonucleic acid polymerase. Biochem. J. 129, 1167 (1972).

DAVIES, P., GRIFFITHS, K.: Stimulation *in vitro* of prostatic ribonucleic acid polymerase by 5α-dihydrotestosterone receptor complexes. Biochem. biophys. Res. Commun. 53, 373 (1973).

DAVIES, P., GRIFFITHS, K.: Stimulation of ribonucleic acid polymerase activity *in vitro* by prostatic steroid-receptor complexes. Biochem. J. 136, 611 (1974).

DAVIES, P., GRIFFITHS, K.: Similarities between 5α-dihydrotestosterone receptor complexes from human and rat prostatic tissue: effects on RNA polymerase activity. Molec. cell. Endocr. 3, 143 (1975).

DAVIES, J., NAFTOLIN, F., RYAN, K. J., SIU, J.: A specific high-affinity, limited capacity estrogen receptor component in the cytosol of human fetal pituitary. J. clin. Endocr. 40, 909 (1975).

DAVIS, J. S., MEYER, R. K., MCSHAN, W. H.: Effect of androgen and estrogen on succinic dehydrogenase and cytochrome oxidase of rat prostate and seminal vesicle. Endocrinology 44, 1 (1949).

DEANE, H. W., WURZELMANN, S.: Electron microscopic observations on the post-natal differentiation of the seminal vesicle epithelium of the laboratory mouse. Amer. J. Anat. 117, 91 (1965).

DENEF, C.: Effect of hypophysectomy and pituitary implants at puberty on the sexual differentiation of testosterone metabolism in rat liver. Endocrinology 94, 1577 (1974).

DENTON, M. J., SPENCER, N., ARNSTEIN, H. R. V.: Biochemical and enzymic changes during erythrocyte differentiation: the significance of the final cell division. Biochem. J. 146, 205 (1975).

DIXIT, V. P., NIEMI, M.: Action of testosterone administered neonatally on the rat perineal complex. J. Endocr. 59, 379 (1973).

DOERING, C. H., KRAEMER, H. C., BRODIE, K. H., HAMBURG, D. A.: A cycle of plasma testosterone in the human male. J. clin. Endocr. 40, 492 (1975).

DORFMAN, R. I., SHIPLEY, R. A.: Androgens. Biochemistry, physiology and clinical significance. New York: John Wiley & Sons 1956.
DORFMAN, R. I., UNGAR, F.: Metabolism of steroid hormones. New York: Academic Press 1965.
DOUGHTY, C., BOOTH, J. E., MCDONALD, P. G., PARROTT, R. F.: The effects of oestradiol-17β, oestradiol benzoate and the synthetic oestrogen RU 2858 on sexual differentiation in the neonatal female rat. J. Endocr. 67, 419 (1975).
DREWS, U., ITAKURA, R., DOFUKU, R., TETTENBORN, U., OHNO, S.: Nuclear DHT-receptor in Tfm/y kidney cell. Nature (Lond.) [New Biol.] 238, 216 (1972).
DUBÉ, J. Y., TREMBLAY, R. R., LESAGE, R., VIERRET, G.: *In vivo* uptake and metabolism by the head appendages of the cock. Molec. cell. Endocr. 2, 213 (1975).
DUNCAN, D., NALL, D., MORALES, R.: Observations on the fine structure of old age pigment. J. Geront. 15, 366 (1960).
EDELMAN, I. S., BOGOROCH, R., PORTER, G. A.: On the mechanism of action of aldosterone on sodium transport: the role of protein synthesis. Proc. nat. Acad. Sci. (Wash.) 50, 1169 (1963).
EINARSSON, K., GUSTAFSSON, J-Å, STENBERG, A.: Neonatal imprinting of liver microsomal hydroxylation and reduction of steroids. J. biol. Chem. 248, 4987 (1973).
EISEN, H. J., GLINSMANN, W.: Partial purification of glucocorticoid receptor from rat liver using DNA-cellulose. J. Steroid Biochem. 6, 1171 (1975).
EPPENBERGER, U., HSIA, S. L.: Binding of steroid hormones by the 105,000 × g supernatant fraction from homogenates of rat skin and variations during the hair cycle. J. biol. Chem. 247, 5463 (1972).
EVANS, C. R., PIERREPOINT, C. G.: Demonstration of a specific cytosol receptor in the normal and hyperplastic canine prostate for 5α-androstane-3α,17α-diol. J. Endocr. 64, 539 (1975).
FANG, S., ANDERSON, K. M., LIAO, S.: Receptor proteins for androgens. On the role of specific proteins in the retention of 17β-hydroxy-5α-androstan-3-one by the rat ventral prostate *in vivo* and *in vitro*. J. biol. Chem. 244, 6584 (1969).
FANG, S., LIAO, S.: Antagonistic action of antiandrogens on the formation of a specific dihydrotestosterone receptor complex in rat ventral prostate. Molec. Pharmacol. 5, 420 (1969).
FANG, S., LIAO, S.: Androgen receptors. Steroid and tissue-specific retention of a 17β-hydroxy-5α-androstane-3-one-protein complex by the cell nuclei of ventral prostate. J. biol. Chem. 246, 16 (1971).
FARNSWORTH, W. E.: A possible mechanism of regulation of rat prostate function by testosterone. Biochim. biophys. Acta 150, 446 (1968).
FARNSWORTH, W. E., BROWN, J. R.: Testosterone metabolism in the prostate. Nat. Cancer Inst. Monogr. 12, 323 (1963).
FEYEL-CABANES, T.: Étude biologique d'un analogue de la méthyl testostérone. C.R. Acad. Sci. (Paris) 157, 1428 (1963).
FICHER, M., STEINBERGER, E.: Conversion of progesterone to testosterone by testicular tissue at different stages of maturation. Steroids 11, 351 (1968).
FISHMAN, W. H.: β-Glucuronidase and the action of steroid hormones. Am. N. Y. Acad. Sci. 54, 548 (1951).
FISHMAN, W. H., IDE, H., RUFO, R.: Dual localization of acid hydrolases in endoplasmic reticulum and in lysosomes. Histochemie 20, 287 (1969).
FLICKINGER, C. J.: The fine structure and development of the seminal vesicles and prostate in the fetal rat. Z. Zellforsch. mikrosk. Anat. 109, 1 (1970 a).
FLICKINGER, C. J.: Extracellular specializations associated with hemidesmosomes in the fetal rat urogenital sinus. Anat. Res. 168, 195 (1970 b).
FLICKINGER, C. J.: Ultrastructural observations on the post-natal development of the rat prostate. Z. Zellforsch. 113, 157 (1971).
FLICKINGER, C. J.: Fine structural aspects of cytodifferentiation. Male accessory sex organs, p. 115, D. Brandes, (ed.). New York: Academic Press 1974.
FOLCA, P. J.: GLASCOCK, R. F., IRVINE, S. T.: Studies with tritium-labelled hexoestrol

in advanced breast cancer. Comparison of tissue accumulation of hexoestrol with bilateral adrenalectomy and oophorectomy. Lancet ii, 796 (1961).

FOLMAN, Y., SOWELL, J. G., EIK-NES, K. B.: The presence and formation of 5α-dihydrotestosterone in rat testis *in vivo* and *in vitro*. Endocrinology **91**, 702 (1972).

FOX, T. O., PARDEE, A. R.: Proteins made in the mammalian cell cycle. J. biol. Chem. **246**, 6159 (1971).

FRANKER, C. K., PRICHARD, C. D., LAMBDEN, C. A.: DNA-binding proteins and the cell cycle in *Cryptothecedonium cohnii*. Differentiation **1**, 383 (1973).

FREDERIKSEN, D. W., WILSON, J. D.: Partial characterization of the nuclear reduced nicotinamide adenine dinucleotide phosphate: Δ^4-3-ketosteroid 5α-oxidoreductase of rat prostate. J. biol. Chem. **246**, 2584 (1971).

FRENCH, F. S., RITZÉN, E. M.: Androgen-binding protein in efferent duct fluid of rat testis. J. Reprod. Fertil. **32**, 479 (1973).

FRENCH, F. S., VAN WYCK, J. J., BAGGETT, B., EASTERLING, W. E., TALBERT, L. M., JOHNSTON, F. R., FORCHIELLI, E., DAY, A. C.: Further evidence of a target organ defect in the syndrome of testicular feminization. J. clin. Endocr. **26**, 493 (1966).

FRENSTER, J. H., ALLFREY, V. G., MIRSKY, A. E.: Repressed and inactive chromatin isolated from interphase lymphocytes. Proc. nat. Acad. Sci. (Wash.) **50**, 1026 (1963).

FRIEND, D. S.: Cytochemical staining of multivesicular body and golgi vesicles. J. Cell Biol. **41**, 269 (1969).

GALLWITZ, D., MUELLER, G. C.: Histone synthesis *in vitro* on HeLa cell microsomes. The nature of the coupling to DNA synthesis. J. biol. Chem. **244**, 5947 (1969).

GANGULY, M., WARREN, J. C.: Affinity labelling of steroid binding sites. Synthesis of cortisone 21-iodoacetate and a study of 20β-hydroxysteroid dehydrogenase. J. biol. Chem. **246**, 364 (1971).

GANSCHOW, R. E.: The genetic control of acid hydrolases. Metabolic conjugation and metabolic hydrolysis, vol. 2, p. 189, W. H. Fishman, (ed.). New York: Academic Press 1973.

GANSCHOW, R. E., BUNKER, B. G.: Genetic control of glucuronidase in mice. Biochem. Genet. **4**, 127 (1970).

GANSCHOW, R. E., PAIGEN, K.: Separate genes determining the structure and intracellular location of hepatic glucuronidase. Proc. nat. Acad. Sci. (Wash.) **58**, 938 (1967).

GANSCHOW, R. E., PAIGEN, K.: Glucuronidase phenotypes of inbred mouse strains. Genetics **59**, 335 (1968).

GATENBY, J. B.: The Golgi apparatus of the living sympathetic ganglion cell of the mouse, photographed by phase-contrast microscopy. J. roy. micr. Soc. **73**, 61 (1953).

GEHRING, U., TOMKINS, G. M., OHNO, S.: Effect of the androgen-insensitivity mutation on a cytoplasmic receptor for dihydrotestosterone. Nature (Lond.) [New Biol.] **232**, 106 (1971).

GELLER, J., VAN DAMME, O., GARABIETA, G., LOH, A., RETTURA, J., SEIFTER, E.: Effect of cyproterone acetate on the [^{3}H] testosterone uptake and enzyme synthesis by ventral prostate of the rat. Endocrinology **84**, 1330 (1969).

GERRALL, A. A., MCMURRAY, M. M., FARRELL, A.: Suppression of the development of the female hamster by implanting testosterone and non-aromatizable androgens administered neonatally. J. Endocr. **67**, 439 (1975).

GHANADIAN, R., LUMIS, J. G., CHISHOLM, G. D.: Serum testosterone and dihydrotestosterone changes with age in rat. Steroids **25**, 753 (1975).

GHRAF, R., HOFF, H-G., LAX, E. R., SCHRIEFFERS, H.: Enzyme activity in kidney, adrenal and gonadal tissue of rats treated neonatally with androgens or oestrogens. J. Endocr. **67**, 317 (1975).

GIANNOPOULOS, G.: Binding of testosterone to uterine components of the immature rat. J. biol. Chem. **248**, 1004 (1973).

GILBERT, W., MÜLLER-HILL, B.: The lac operator is DNA. Proc. nat. Acad. Sci. (Wash.) **58**, 2415 (1967).

GIORGI, E. P.: Studies on androgen transport into canine prostate *in vitro*. J. Endocr. **68**, 109 (1976).

GIORGI, E. P., MOSES, T. F., GRANT, J. K., SCOTT, R., SINCLAIR, J.: *In vitro* studies on the regulation of androgen-tissue relationships in canine normal and human hyperplastic prostate. Molec. cell. Endocr. 1, 271 (1974).

GIORGI, E. P., SHIRLEY, I. M., GRANT, J. K., STEWART, J. C.: Androgen dynamics *in vitro* in the human prostate gland. Effect of cyproterone and cyproterone acetate. Biochem. J. 132, 465 (1973).

GITTINGER, J. W., LASNITZKI, I.: The effect of testosterone and testosterone metabolites on the fine structure of the rat prostate gland in organ culture. J. Endocr. 52, 459 (1972).

GIVNER, M. L., JAGARINEC, N.: Inhibition of prostatic dihydrotestosterone formation by medrogestone. J. Steroid Biochem. 5, 352 (1974).

GLOYNA, R. E., SIITERI, P. K., WILSON, J. D.: Dihydrotestosterone in prostatic hypertrophy. II. The formation and content of dihydrotestosterone in the hypertrophic canine prostate and the effect of dihydrotestosterone on prostate growth in the dog. J. clin. Invest. 49, 1746 (1970).

GLOYNA, R. E., WILSON, J. D.: A comparative study of the conversion of testosterone to 17β-hydroxy-5α-androstan-3-one (DHT) by prostate and epididymis. J. clin. Endocr. 29, 970 (1969).

GOLDMAN, R. D., POLLACK, R., HOPKINS, N. H.: Preservation of normal behaviour by enucleated cells in culture. Proc. nat. Acad. Sci. (Wash.) 70, 750 (1973).

GOLDSTEIN, J. L., WILSON, J. D.: Genetic and hormonal control of male sexual differentiation. J. cell. Physiol. 85, 365 (1975).

GOMEZ, E. C., HSIA, S. L.: *In vitro* metabolism of testosterone-4-^{14}C and Δ^4-androstene-3,17-dione-4-^{14}C in human skin. Biochemistry 7, 24 (1968).

GONZALEZ-DIDDI, M., KOMISAZUK, B., BEYER, C.: Differential effects of testosterone and dihydrotestosterone on the diverse uterine tissue of the ovariectomized rat. Endocrinology 91, 1129 (1972).

GORDON, G. G., ALTMAN, K., SOUTHREN, A. L., OLIVO, J.: Human hepatic testosterone A-ring reductase activity: effect of medroxyprogesterone acetate. J. clin. Endocr. 32, 457 (1971).

GORDON, J., SMITH, J. A., KING, R. J. B.: Metabolism and binding of androgens by mouse mammary tumour cells in culture. Molec. cell. Endocr. 1, 259 (1974).

GORSKI, J., RAKER, B.: The effects of cytochalasin B on estrogen binding and 2-deoxyglucose metabolism in the rat uterus. Endocrinology 93, 1212 (1973).

GORSKI, R. A.: Modulations of ovulatory mechanisms by postnatal administration of oestrogen to the rat. Amer. J. Physiol. 205, 842 (1963).

GOTTLEIB, H., GERRALL, A. A., THIEL, A. R.: Receptivity in female hamsters following neonatal testosterone propionate and MER-25. Physiol. Behav. 12, 61 (1974).

GRACY, R. W., LACKO, A. G., BROK, L. W., ADELMAN, R. C., HORECKER, B. L.: Structural relations in aldolases purified from rat liver and muscle and Novikoff hepatoma. Arch. Biochem. 136, 480 (1970).

GRACY, R. W., LACKO, A. G., HORECKER, B. L.: Subunit structure and chemical properties of rabbit liver aldolase. J. biol. Chem. 244, 3913 (1969).

GRANICK, S., KAPPAS, A.: Steroid induction of porphyrin synthesis in liver cell culture. J. biol. Chem. 242, 4587 (1967).

GRAYHACK, J. T., LEBOVITZ, J. M.: Effect of prolactin on the citric acid and lateral lobe of the prostate of Sprague-Dawley rats. Invest. Urol. 5, 87 (1967).

GREENE, R. R., BURRILL, M. W., THOMSON, D. M.: Further studies on the androgenicity of progesterone. Endocrinology 27, 469 (1940).

GROSS, S. R., ARONOW, L., PRATT, W. B.: The active transport of cortisol by mouse fibroblasts growing *in vitro*. Biochem. biophys. Res. Commun. 32, 66 (1968).

GROSS, S. R., ARONOW, L., PRATT, W. B.: The outward transport of cortisol by mammalian cells *in vitro*. J. Cell Biol. 44, 103 (1970).

GROSSMAN, S. H., AXELROD, B., BARDIN, C. W.: Effect of testosterone on renal and hepatic L-gluconolactonase activities in male, female and pseudohermaphrodite rats. Life Sci. 10, 175 (1971).

GROVER, P. K., O'DELL, W. D.: Correlation of *in vivo* and *in vitro* activities of some naturally occurring androgens using a radio-receptor assay for 5α-dihydrotestosterone with rat prostate cytosolic receptor protein. J. Steroid Biochem. **6**, 1373 (1975).

GUDIS, S., HYODO, M., EBERLE, H.: Identification of a protein from *Escherichia coli* whose synthesis appears to be triggered by the initiation of DNA replication. Biochem. biophys. Res. Commun. **62**, 1003 (1975).

GUÉRIGUAN, J. L., PEARLMAN, W. H.: Some properties of a testosterone binding component of human pregnancy serum. J. biol. Chem. **243**, 5226 (1968).

GUILLEMANT, S., CORVOL, P., CRÉPY, O., MAUVAIS-JERVIS, P.: Action du cyprotérone sur la liaison plasmatique et le métabolisme de la testostérone. C.R. Acad. Sci. (Paris) **268**, 2595 (1969).

GURDON, J. B., WOODLAND, M. R.: The cytoplasmic control of nuclear activity in animal development. Biol. Rev. **43**, 233 (1968).

GURPIDE, E., WELCH, M.: Dynamics of uptake of estrogens and androgens by human endometrium. J. biol. Chem. **244**, 5159 (1969).

GUSTAFSSON, J-Å., INGELMAN-STUNDBERG, M., STENBERG, Å., NEUMANN, F.: Partial feminization of hepatic steroid metabolism in male rats after the neonatal administration of cyproterone acetate. J. Endocr. **64**, 267 (1975 a).

GUSTAFSSON, J-Å., POUSETTE, Å., STENBERG, Å., WRANGE, Ö.: High affinity binding of 4-androstene-3,17-dione in rat liver. Biochemistry **14**, 3942 (1975 b).

GUSTAFSSON, J-Å., STENBERG, Å.: Masculinization of rat liver enzymes following hypophysectomy. Endocrinology **95**, 891 (1974 a).

GUSTAFSSON, J-Å., STENBERG, Å.: Influence of adrenal steroids on liver enzymes of neonatally castrated rats. J. Endocr. **63**, 103 (1974 b).

HAGENAS, L., RITZÉN, E. M.: Impaired Sertoli cell function in experimental cryptorchidism in the rat. Molec. cell. Endocr. **4**, 25 (1976).

HAINES, M. E., HOLMES, A. M., JOHNSTON, I. R.: Distinct cytoplasmic and nuclear DNA polymerases from rat liver. FEBS Lett. **17**, 63 (1971).

HAINES, M. E., WICKREMASINGHE, R. G., JOHNSTON, I. R.: Purification and partial characterization of rat-liver nuclear DNA polymerase. Europ. J. Biochem. **31**, 119 (1972).

HÄKÖNEN, P., ISOTALO, A., SANTTI, R.: Studies on the mechanism of testosterone action on glucose metabolism in the rat ventral prostate. J. Steroid Biochem. **6**, 1405 (1975).

HANSSON, V.: Further characterization of the 5α-dihydrotestosterone binding in the epididymal cytosol fraction. *In vitro* studies. Steroids **20**, 575 (1972).

HANSSON, V., DJOSELAND, O., REUSCH, E., ATTRAMADAL, A., TORGESEN, O.: An androgen-binding protein in the testis cytosol fraction of adult rats. Comparison with the androgen-binding protein in the epididymis. Steroids **21**, 457 (1973).

HANSSON, V., MCLEAN, W. S., SMITH, A. A., TINDALL, D. J., WEDDINGTON, S. C., NAYFEH, S. N., FRENCH, F. S., RITZÉN, E. M.: Androgen receptors in rat testis. Steroids **23**, 823 (1974).

HANSSON, V., TVETER, K. J., ATTRAMADAL, A., TORGESEN, O.: Androgenic receptors in human benign nodular prostatic hyperplasia. Acta endocr. (Kbh.) **68**, 79 (1971).

HANSSON, V., TVETER, K. J., UNHJEM, O., DJØSELAND, O., ATTRAMADAL, A., REUSCH, E., TORGESEN, O.: Androgen binding in male sex organs with special reference to the human prostate. Normal and abnormal growth of the prostate, p. 676, M. Goland, (ed.). Springfield, Ill.: C. C. Thomas and Co. 1975.

HARDING, B. W., SAMUELS, L. T.: A tissue fractionation study of rat ventral prostate: subcellular distribution of nucleic acids, succinate oxidising systems, cytochrome oxidase, cytochrome c reductase and acid phosphatase. Biochim. biophys. Acta, **54**, 42 (1961).

HARNS, G. W., DONOVAN, B. T.: The pituitary gland. London: Butterworth 1966.

HARPER, M. E., PIERREPOINT, C. G., FAHMY, A. R., GRIFFITHS, K.: The effects of prostatic metabolites of testosterone and other substances on the isolated deoxyribonucleic acid polymerase of the canine prostate. Biochem. J. **119**, 785 (1970).

HARPER, M. E., PIERREPOINT, C. G., FAHMY, A. R., GRIFFITHS, K.: The metabolism of steroids in the canine prostate and testis. J. Endocr. **49**, 213 (1971).

HARRISON, R. W., FAIRFIELD, S., ORTH, D. N.: Evidence for glucocorticoid transport into ACT-20/DI cells. Biochemistry **14**, 1304 (1975).

HAYES, K. J.: The so-called "levator ani" of the rat. Acta endocr. (Kbh.) **48**, 337 (1965).

HAYTON, G. J., PEARSON, C. K., SCAIFE, J. R., KEIR, H. M.: Synthesis of DNA in BHK-21/C13 cells. Biochem. J. **131**, 499 (1973).

HEAP, R. B., SYMONS, A. M., WATKINS, J. C.: Steroids and their interaction with phospholipids: solubility, distribution coefficient and effect on potassium permeability of liposomes. Biochim. biophys. Acta **218**, 482 (1970).

HECHT, W. S.: Enzymatically active intermediate in the conversion between the low and high molecular weight DNA polymerases. Nature (Lond.) [New Biol.] **245**, 199 (1973).

HEINRICHS, W. L., KARZNIA, R., WYSS, R., HERMANN, W. L.: Testicular ferminization and apparent enzyme defect. Clin. Res. **17**, 143 (1973).

HEINZEL, W., RAHIMI-LARIDJANI, I., GRIMMINGER, H.: Immunoadsorbents: non-specific binding of proteins to albumin-Sepharose. J. immunol. Methods **9**, 337 (1976).

HELDE, R. L. VAN DE, VAN DE HELDE, S. C.: Annulated lamellae in the ductus epididymis of fetal and castrated adult rats. Anat. Rec. **161**, 427 (1968).

HELMINEN, H. J., ERICSSON, J. L. E.: On the mechanism of lysosomal enzyme secretion. Electron microscopic and histochemical studies on the epithelial cells of the rat's ventral prostate lobe. J. Ultrastruct. Res. **33**, 528 (1970).

HEWISH, D. R., BURGOYNE, L. A.: Chromatin substructure. The digestion of chromatin DNA at regularly spaced sites by nuclear deoxyribonuclease. Biochem. biophys. Res. Commun. **52**, 504 (1973).

HEYDEN, H. W., ZACHAU, H.: Characteristics of RNA in fractions of calf thymus chromatin. Biochim. biophys. Acta **232**, 651 (1971).

HIGGINS, S. J., BURCHELL, J. M., MAINWARING, W. I. P.: Androgen-dependent synthesis of basic secretory proteins by the rat seminal vesicle. Biochem. J.: **158**, 271 (1976).

HIGGINS, S. J., ROUSSEAU, G. G., BAXTER, J. D., TOMKINS, G. M.: Early events in glucocorticoid action. Activation of the steroid receptor and its subsequent specific nuclear binding studied in a cell-free system. J. biol. Chem. **248**, 5866 (1973 a).

HIGGINS, S. J., ROUSSEAU, G. G., BAXTER, J. D., TOMKINS, G. M.: Nuclear binding of steroid receptors: comparison in intact cells and cell-free systems. Proc. nat. Acad. Sci. (Wash.) **70**, 3415 (1973 b).

HIGUCHI, T., VILLEE, C. A.: Aromatization of epitestosterone by human placenta. Endocrinology **86**, 912 (1970).

HIMMELWEIT, F.: The collected papers of Paul Ehrlich, Volume 3. London: Academic Press 1960.

HINKLE, D. C., CHAMBERLIN, M.: Studies on the binding of *Escherichia coli* RNA polymerase to DNA. 1. The role of sigma subunit in site selection. J. molec. Biol. **70**, 157 (1972).

HIPPEL, P. H. VON, MCGHEE, J.: DNA protein interactions. Ann. Rev. Biochem. **41**, 231 (1972).

HIPPEL, P. H. VON, REVZIA, A., GROSS, C. A., WANG, A. C.: Non-specific DNA binding of genome regulating proteins as a biological control mechanism. 1. The lac operon: equilibrium aspects. Proc. nat. Acad. Sci. (Wash.) **71**, 4808 (1974).

HNILICA, L. S.: The structure and biological function of histones. Cleveland: Chemical Rubber Company Press, 1972.

HØISAETER, P. A.: The androgen cytosol receptor in the ventral prostate of the rat. Methodological problems. Biochim. biophys. Acta **317**, 492 (1973).

HØISAETER, P. A.: Studies on a hormone-cytostatic complex in the cytoplasm of rat prostate gland. Invest. Urol. **12**, 33 (1974).

HOLMES, D. S., MAYFIELD, J. E., SANDER, G., BONNER, J.: Chromosomal RNA: its properties. Science **177**, 72 (1972).

HONMA, Y., NOUMURA, T.: Androgen regulation of rapidly synthesized RNA in the rat ventral prostate. J. Endocr. **65**, 377 (1975).

HORN, H., FINKELSTEIN, M.: Aromatization of epitestosterone by human placenta. Endocrinology **88**, 271 (1971).

HORTON, R., KATO, T., SHERINS, R.: A rapid method for the estimation of testosterone in male plasma. Steroids **10**, 245 (1967).

HORTON, R., TAIT, J. F.: Androstenedione production and interconversion rates measured in peripheral blood and studies on the possible site of its conversion to testosterone. J. clin. Invest. **45**, 301 (1966).

HOWARD, A., PELC, S. R.: Nuclear incorporation of P^{32} as demonstrated by autoradiography. Exp. Cell Res. **2**, 178 (1951).

HUBERMAN, J. A., KORNBERG, A., ALBERTS, B. M.: Stimulation of T4 bacteriophage DNA polymerase by the protein product of gene 32. J. molec. Biol. **62**, 39 (1971).

HUGGINS, C., HODGES, C. V.: Studies on prostatic cancer. I. The effect of castration and estrogen and of androgen injection on serum phosphatasis in metastatic carcinoma of the prostate. Cancer Res. **1**, 293 (1943).

HUGGINS, C., JENSEN, E. V., CLEVELAND, A. S.: Chemical structure of steroids in relation to promotion of growth of the vagina and uterus of the hypophysectomized rat. J. exp. Med. **100**, 225 (1954).

HUGGINS, C., MAINZER, K.: Hormonal influences on mammary tumors of the rat. J. exp. Med. **105**, 485 (1957).

HUMPHREY, G. F., MANN, T.: Studies on metabolism of semen. 5. Citric acid in semen. Biochem. J. **44**, 97 (1949).

HUNT, W. L., NICHOLSON, N.: Antiandrogenic activities of 17β-hydroxy-16,16,dimethylestr-4-en-3-one (SC-14207) on sebum and hamster flank organs. J. Steroid Biochem. **5**, 352 (1974).

ICHII, S., IZAWA, M., MURAKAMI, N.: Hormonal regulation of protein synthesis in rat ventral prostate. Endocr. jap. **21**, 267 (1974).

IDLER, D. R., FREEMAN, H. C.: Binding of testosterone, 1α-hydroxycorticosterone and cortisol by plasma proteins of fish. Gen. comp. Endocr. **11**, 366 (1968).

IKEHARA, Y., ENDO, H., OKADA, Y.: The identity of the aldolases isolated from rat muscle and primary hepatoma. Arch. Biochem. **136**, 491 (1970).

INANO, H., YORI, Y., TAMAOKI, B.-I.: Effect of age on testicular enzymes related to steroid bioconversion. Ciba Found. Colloq. Endocr. **16**, 105 (1967).

IRVING, R. A., MAINWARING, W. I. P.: Partial purification of steroid-receptor complexes by DNA-cellulose chromatography and isoelectric focusing. J. Steroid Biochem. **5**, 711 (1974).

IRVING, R. A., MAINWARING, W. I. P., SPOONER, P. M.: The regulation of haemoglobin synthesis in cultured chick blastoderms by steroids related to 5β-androstane. Biochem. J. **154**, 83 (1976).

ITO, T., HORTON, R.: Dihydrotestosterone in human peripheral plasma. J. clin. Endocr. **31**, 362 (1970).

ITO, T., HORTON, R.: The source of plasma dihydrotestosterone in man. J. clin. Invest. **50**, 1621 (1971).

IVARIE, R. D., FAN, W.J-W., TOMKINS, G. M.: Analysis of the induction and deinduction of tyrosine aminotransferase in enucleated HTC cells. J. cell. Physiol. **85**, 357 (1975).

JACOB, S. T.: Mammalian RNA polymerases. Progr. nucleid Acid Res. **13**, 93 (1973).

JACOB, S. T., JÄNNE, O., SAJDEL-SULKOWSKA, E. M.: Hormonal regulation of RNA polymerases in rat liver and kidney. Isoenzymes, vol. III, p. 9, C. L. Maskert (ed.). New York: Academic Press 1975.

JENDRISAK, J. J., BURGESS, R. R.: A new method for the large-scale preparation of wheat germ DNA-dependent RNA polymerase II. Biochemistry **14**, 4639 (1975).

JENSEN, E. V., JACOBSEN, H. I.: Basic guides to the mechanism of estrogen action. Recent Progr. Horm. Res. **18**, 387 (1962).

JENSEN, E. V., SUSUKI, T., KAWASHIMA, T., STUMPF, W. E., DE SOMBRE, E. R.: A two step mechanism for the interaction of estradiol with rat uterus. Proc. nat. Acad. Sci. (Wash.) 59, 632 (1968).

JOHNSON, J. D., ST. JOHN, T., BONNER, J.: DNA-binding proteins from Novikoff hepatoma cells. Biochim. biophys. Acta 378, 424 (1974).

JOSHI, M. S., YARON, A., LINDNER, H. R.: Intrauterine gelation of seminal plasma components in the rat after coitus. J. Reprod. Fertil. 30, 27 (1972).

JOSSO, N.: Interspecific character of the Müllerian-inhibiting substance: action of the human fetal testis, ovary and adrenal on the fetal rat Müllerian duct in organ culture. J. clin. Endocr. 32, 404 (1971).

JOSSO, N.: Permeability of membranes to the Müllerian-inhibiting substance synthesized by the human fetal testis *in vitro:* a clue to its biochemical nature. J. clin. Endocr. 34, 265 (1972).

JOSSO, N.: *In vitro* synthesis of Müllerian-inhibiting hormone by seminiferous tubules isolated from calf fetal testis. Endocrinology 93, 829 (1974).

JOST, A.: The problems of fetal endocrinology; the gonadal and hypophysial hormones. Recent Progr. Horm. Res. 8, 379 (1953).

JOST, A.: The role of fetal hormones in prenatal development. Harvey Lect. 55, 201 (1959).

JOST, A.: Steroids and sex differentiation of the mammalian foetus. Excerpta Medica Int. Congr. Ser. 132, 74 (1967).

JOST, A.: Hormonal factors in the sex differentiation of the mammalian foetus. Phil. Trans. B. 259, 119 (1970).

JOUAN, P., SAMPAREZ, S., THIEULANT, M. L.: Testosterone receptors in purified nuclei of rat anterior hypophysis. J. Steroid Biochem. 4, 65 (1973).

JUNG, I., BAULIEU, E-E.: Testosterone cytosol "receptor" in the rat levator ani muscle. Nature (Lond.) [New Biol.] 237, 24 (1972).

KALIMI, M., COLMAN, P., FIEGELSON, P.: The "activated" hepatic glucocorticoid receptor complex. Its generation and properties. J. biol. Chem. 250, 1080 (1975).

KALIMI, M., TSAI, S. Y., TSAI, M-J., CLARK, J. H., O'MALLEY, B. W.: Effect of estrogen on gene expression in chick oviduct. Correlation between nuclear-bound estrogen receptor and chromatin initiation sites for transcription. J. biol. Chem. 251, 516 (1976).

KARL, T. R., CHAPMAN, V. M.: Linkage and expression of the Eg locus control by inclusion of β-glucuronidase into microsomes. Biochem. Genet. 11, 367 (1974).

KARSZIA, R., WYSS, R. H., HEINRICHS, W. R., HERMANN, W. L.: Binding of pregnenolone and progesterone by prostatic "receptor" proteins. Endocrinology 84, 1238 (1969).

KATCHEN, B., BUXBAUM, S.: Disposition of a new, nonsteroid antiandrogen, α,α,α-trifluoro-2-methyl-4'-nitro-m-propionotoluidide (flutamide) in men following a single oral 200 mg dose. J. Clin. Endocr. 41, 373 (1975).

KATO, J.: Localization of oestradiol receptors in the rat hypothalamus. Acta endocr. (Kbh.) 72, 663 (1973).

KATO, J., ATSUMI, Y., INABA, M.: Estradiol receptors in female rat hypothalamus in the developmental stages and during pubescence. Endocrinology 94, 309 (1974).

KATO, T., HORTON, R.: Studies on testosterone binding globulin. J. clin. Endocr. 28, 1160 (1968).

KATZENELLENBOGEN, B. S., GORSKI, J.: Estrogen action *in vitro*. Induction of the synthesis of a specific uterine protein. J. biol. Chem. 247, 1299 (1972).

KATZENELLENBOGEN, J., JOHNSON, H. J., CARLSON, K. E., MYERS, H. N.: Photoreactivity of some light-sensitive estrogen derivatives. Use of an exchange assay to determine the photointeraction with the rat uterine estrogen binding protein. Biochemistry 13, 2986 (1974).

KEENAN, B. S., MEYER, W. T., HADJIAN, A. J., MIGEON, C. J.: Androgen receptor in skin fibroblasts: characterization of a specific 17β-hydroxy-5α-androstane-3-one-protein complex in cell sonicates and nuclei. Steroids 25, 535 (1975).

KEIR, H. M., CRAIG, R. K.: The regulation of deoxyribonucleic acid synthesis. Biochem. Soc. Trans. 1, 1073 (1973).

KEIR, L. B.: Molecular orbital theory in drug research. New York: Academic Press 1971.

KELCH, R. P., LINDHOLM, U. B., JAFFE, R. B.: Testosterone metabolism in target tissues: 2. Human fetal and adult reproductive tissues, perineal skin and skeletal muscle. J. clin. Endocr. 32, 449 (1971).

KELLER, W.: Characterization of purified DNA-relaxing enzyme from human tissue culture cells. Proc. nat. Acad. Sci. (Wash.) 72, 2550 (1975).

KENT, J. R., ACONE, A. B.: Plasma testosterone and ageing in males. Androgens in normal and pathological conditions, p. 31, A. Vermeulen and D. Exley (eds.). Amsterdam: Excerpta Medica Foundation 1966.

KESHGEGIAN, A. A., FURTH, J. J.: Comparison of transcription of chromatin by calf thymus and *E. coli* RNA polymerase. Biochem. biophys. Res. Commun. 48, 757 (1972).

KING, R. J. B., GORDON, J.: Involvement of DNA in the acceptor mechanism for uterine oestradiol receptor. Nature (Lond.) [New Biol.] 240, 185 (1972).

KING, R. J. B., MAINWARING, W. I. P.: Steroid-Cell Interactions. London: Butterworths 1974.

KIRSCHNER, M. A., COHEN, F. B., JESPERSEN, D.: Estrogen production and its origin in men with gonadotrophin-producing neoplasms. J. clin. Endocr. 39, 112 (1974).

KISH, V. M., KLEINSMITH, L. J.: Nuclear protein kinases. Evidence for their heterogeneity, tissue specificity, substrate specificities, and differential responses to adenosine 3′:5′-monophosphate. J. biol. Chem. 249, 750 (1974).

KIT, S., LEUNG, W-C., TRKULA, D., JORGENSEN, G.: Gel electrophoresis and isoelectric focusing of mitochondrial and viral thymidine kinases. Int. J. Cancer. 13, 203 (1974).

KITAY, J.: Effects of estrogen and androgen on the adrenal cortex of the rat. Functions of the adrenal cortex, vol. 2, p. 775, K. W. McKerns (ed.). New York: Appleton Century Crofts, 1968.

KLEINSMITH, L. J.: Specific binding of phosphorylated non-histone chromatin proteins to deoxyribonucleic acid. J. biol. Chem. 248, 5648 (1973).

KLEINSMITH, L. J., HEIDEMA, J., CARROLL, A.: Specific binding of rat liver nuclear proteins to DNA. Nature (Lond.) 226, 1025 (1970).

DE KLOET, R., WALLACH, G., MCEWEN, B. S.: Differences in corticosterone and dexamethasone binding to rat brain and pituitary. Endocrinology 96, 598 (1975).

KNAPSTEIN, P., AMNON, D., WU, C. H., ARCHER, D. F., FLICKERING, G. L., TOUCHSTONE, J. C.: Metabolism of free and sulfoconjugated DHEA in brain tissue *in vivo* and *in* vitro. Steroids 11, 885 (1968).

KOCHAKIAN, C. D.: Definition of androgens and protein anabolic steroids. Pharmacol. ther. Biol. 1, 149 (1975).

KONTULA, K., JÄNNE, O., VIHKO, R., DE JAGER, E., DE VISSER, J., ZEELEN, F.: Progesterone-binding proteins: *in vitro* binding and biological activity of different steroid ligands. Acta endocr. (Kbh.) 78, 574 (1975).

KORACH, K. S., MULDOON, T. G.: Studies on the nature of the hypothalamic estradiol concentrating mechanism in the male and female rat. Endocrinology 94, 785 (1974).

KORACH, K. S., MULDOON, T. G.: Inhibition of anterior pituitary estrogen-receptor complex formation by low-affinity interaction with 5α-dihydrotestosterone. Endocrinology 97, 231 (1975).

KORENBROT, C. C., PAUP, D. C., GORSKI, R. A.: Effects of testosterone propionate and dihydrotestosterone propionate on plasma FSH and LH levels in neonatal rats and on sexual differentiation of the brain. Endocrinology 97, 709 (1975).

KORNBERG, R. D.: Chromatin structure: a repeating unit of histones and DNA. Science 184, 868 (1974).

KORNBERG, R. D, THOMAS, J. O.: Chromatin structure: oligomers of the histones. Science 184, 865 (1974).

KOWARSKI, A., SHALF, J., MIGEON, C. J.: Concentration of testosterone and dihydrotestosterone in subcellular fractions of liver, prostate and muscle in the male dog. J. biol. Chem. **244,** 5269 (1969).

KRAULIS, I., CLAYTON, R. B.: Sexual differentiation of testosterone metabolism exemplified by the accumulation of 3β, 17α-dihydroxy-5α-androstane-3-sulphate as a metabolite of testosterone in the castrated rat. J. biol. Chem. **234,** 3546 (1968).

KREIG, M., HORST, H-J., STERBA, M-L.: Binding and metabolism of 5α-androstane-3α, 17β-diol and 5α-androstane-3β, 17β-diol in prostate, seminal vesicle and plasma of male rats; studies *in vivo* and *in vitro.* J. Endocr. **64,** 529 (1975).

KREIG, M., SZALAY, R., VOIGT, K. D.: Binding and metabolism of testosterone and 5α-dihydrotestosterone in bulbocavernosus/levator ani (BCLA) of male rats *in vivo* and *in vitro.* J. Steroid Biochem. **5,** 453 (1974).

KUHN, R., SCHRADER, W. T., SMITH, R. G., O'MALLEY, B. W.: Progesterone binding components of chick oviduct. X. Purification by affinity chromatography. J. biol. Chem. **250,** 4220 (1975).

LACROIX, E., EECHAUTE, W., LEUSEN, I.: Influence of age on the formation of 5α-androstanediol and 7α-hydroxytestosterone by incubated rat testis. Steroids **25,** 649 (1975).

LACY, E., AXEL, R.: Analysis of DNA of isolated chromatin subunits. Proc. nat. Acad. Sci. (Wash.) 72, 3978 (1975).

LALLEY, P., SHOWS, T. B.: Lysosomal and microsomal glucuronidase: genetic variants alter electrophoretic mobility of both hydrolases. Science **185,** 442 (1974).

LAN, N. C., KATZENELLENBOGEN, B. S.: Temporal relationships between hormone receptor binding and biological responses in the uterus: studies with short- and long-acting derivatives of estriol. Endocrinology **98,** 220 (1976).

LANE, S. E., GIDARI, A. S., LEVERE, R. D.: Cytoplasmic receptor protein for etiocholanolone in chick embryo liver. J. biol. Chem. **250,** 8209 (1975).

LANGMORE, J. P., WOOLEY, J. C.: Chromatin architecture: investigation of a subunit of chromatin by dark field electron microscopy. Proc. nat. Acad. Sci. (Wash.) 72, 2691 (1975).

LASNITZKI, I., FRANKLIN, H. R.: The influence of serum on the uptake, conversion and action of dihydrotestosterone in rat prostate glands in organ culture. J. Endocr. **64,** 289 (1975).

LAW, L. W., MORROW, A. G., GREENSPAN, E. M.: Inheritance of low liver glucuronidase activity in the mouse. J. nat. Cancer Inst. **12,** 909 (1952).

LAWRENCE, A. M., LANDAU, R. L.: Impaired ventral prostate affinity for testosterone in hypophysectomized rats. Endocrinology **77,** 1119 (1965).

LAZARUS, L. H., KITRON, N.: Cytoplasmic DNA polymerase: polymeric forms and their conversion into an active monomer resembling nuclear DNA polymerase. J. molec. Biol. **81,** 529 (1973).

LEAV, I., MORFIN, R. F., OFNER, P., CAVAZOS, L. F., LEEDS, E. B.: Estrogen- and castration-induced effects on canine prostatic fine structure and C_{19}-steroid metabolism. Endocrinology **89,** 465 (1971).

LEE, A. E.: The effect of continuous oestrogen on mitosis and [^{3}H] thymidine incorporation in the mouse uterus. Basic actions of sex steroids on target cells, p. 243, P. O. Hubinont, F. Leroy and P. Galand (eds.). Basel, S. Karger 1971.

LEHMAN, A. R.: Deoxyribonucleases: their relationship to deoxyribonucleic acid synthesis. Ann. Rev. Biochem. **36,** 645 (1967).

LEVERE, R. D., GRANICK, S.: Control of haemoglobin synthesis in the cultured chick blastoderm by δ-aminolevulinic acid synthetase: increase in the rate of haemoglobin formation with δ-aminolevulinic acid. Proc. nat. Acad. Sci. (Wash.) **54,** 134 (1965).

LEVERE, R. D., KAPPAS, A., GRANICK, S.: Stimulation of haemoglobin synthesis in chick blastoderm by certain 5β-androstane and 5β-pregnane steroids. Proc. nat. Acad. Sci. (Wash.) **58,** 985 (1967).

LEVY, S., SIMPSON, R. T., SOBER, H. A.: Fractionation of chromatin components. Biochemistry **11,** 1547 (1972).

LEWIN, B.: Units of transcription and translation: the relationship between heterogeneous nuclear RNA and messenger RNA. Cell **4**, 11 (1975).

LIANG, T., LIAO, S.: Dihydrotestosterone and the initiation of protein synthesis by prostate ribosomes. J. Steroid Biochem. **6**, 549 (1975a).

LIANG, T., LIAO, S.: A very rapid effect of androgen on initiation of protein synthesis in target tissues. Proc. nat. Acad. Sci. (Wash.) **72**, 706 (1975b).

LIAO, S., BARTON, R. W., LIN, A. H.: Differential synthesis of ribonucleic acid in prostatic nuclei: evidence for selective gene transcription induced by androgens. Proc. nat. Acad. Sci. (Wash.) **55**, 1593 (1966).

LIAO, S., FANG, S.: Receptor-proteins for androgens and the mode of action of androgens on gene transcription in ventral prostrate. Vitam. and Horm. **27**, 17 (1969).

LIAO, S., HOWELL, D. K., CHANG, T-M.: Action of a nonsteroidal antiandrogen, flutamide, on the receptor binding and nuclear retention of 5α-dihydrotestosterone in rat prostate. Endocrinology **94**, 1205 (1974).

LIAO, S., LIANG, T., FANG, S., CASTEÑADA, E., SHAO, T-C.: Steroid structure and androgenic activity. Specificities involved in the receptor binding and nuclear retention of various androgens. J. biol. Chem. **248**, 6154 (1973a).

LIAO, S., LIANG, T., SHAO, T-C., TYMOCZKO, J. L.: Androgen receptor cycling in prostate cells. Advanc. exp. Med. Biol. **36**, 232 (1973 b).

LIAO, S., LIANG, T., TYMOCZKO, J. L.: Structural recognition in interactions of androgens and receptor proteins and in their association with nuclear components. J. Steroid Biochem. **3**, 401 (1972).

LIAO, S., LIANG, T., TYMOCZKO, J. L.: Ribonucleoprotein binding of steroid-"receptor" complexes. Nature (Lond.) [New Biol.] **241**, 211 (1973 c).

LIAO, S., LIN, A. H.: Prostatic nuclear chromatin: an effect of testosterone on the synthesis of RNA rich in cytidylyl (3′,5′) guanosine. Proc. nat. Acad, Sci. (Wash.) **57**, 379 (1967).

LIAO, S., TYMOCZKO, J. L., LIANG, T., ANDERSON, K. M., FANG, S.: Androgen receptors: 17β-hydroxy-5α-androstane-3-one and the translocation of a cytoplasmic protein to cell nuclei in prostate. Adv. Biosci. **7**, 155 (1971).

LIAO, S., WILLIAMS-ASHMAN, H. G.: An effect of testosterone on amino acid incorporation by prostatic ribonucleoprotein particles. Proc. nat. Acad. Sci. (Wash.) **48**, 1956 (1962).

LIEBERBURG, I., MCEWEN, B. S.: Estradiol-17β: a metabolite of testosterone recovered in cell nuclei from limbic areas of neonatal rat brains. Brain Res. **85**, 165 (1975).

LILLIE, F. R.: The free-martin: a study of the action of sex hormones in the foetal life of cattle. J. exp. Zool. **23**, 371 (1917).

LIN, S., RIGGS, A. D.: Lac repressor binding to non-operator DNA: detailed studies and a comparison of equilibrium and rate competition methods. J. molec. Biol. **72**, 671 (1972).

LIPSETT, M. B., TULLNER, W. W.: Testosterone synthesis by the fetal rabbit gonad. Endocrinology **77**, 273 (1965).

LISSER, R. H., CURTIS, L. E., ESCAMILLA, R. F., GOLDBERG, M. B.: The syndrome of congenital aplastic ovaries with sexual infantilism, high urinary gonadotrophins, short stature and other congenital abnormalities. J. clin. Endocr. **7**, 665 (1947).

LITTAU, V. C., BURDICK, C. J., ALLFREY, V. G., MIRSKY, A. E.: The role of histones in the maintenance of chromatin structure. Proc. nat. Acad. Sci. (Wash.) **54**, 1204 (1965).

LONGCOPE, C., KATO, T., HORTON, R.: Conversion of blood androgens to oestrogens in normal adult men and women. J. clin. Invest. **48**, 2191 (1969).

LOUVET, J-P., HARMAN, S. M., SCHREIBER, J. R., ROSS, G. T.: Evidence for a role of androgens in follicular maturation. Endocrinologv **97**, 361 (1975).

LUCAS-LENARD, J., LIPMANN, F.: Protein biosynthesis. Ann. Rev. Biochem. **40**, 409 (1971).

LUTTGE, W. G., WHALEN, R. E.: Dihydrotestosterone, androstenedione, testosterone:

competitive effectiveness in masculinizing and defeminizing reproductive systems in male and female rats. Horm. Behav. 1, 265 (1970).
LYON, M. F., HAWKES, S. G.: X-linked gene for testicular feminisation in the mouse. Nature (Lond.) 227, 1217 (1970).
MAINWARING, W. I. P.: The ageing process in the mouse ventral prostate gland: a preliminary biochemical survey. Gerontologia 13, 177 (1967).
MAINWARING, W. I. P.: Changes in the RNA metabolism of ageing mouse tissues with particular reference to the prostate gland. Biochem. J. 110, 79 (1968).
MAINWARING, W. I. P.: A soluble androgen receptor in the cytoplasm of rat prostate. J. Endocr. 45, 531 (1969 a).
MAINWARING, W. I. P.: The binding of [1,2-^{3}H] testosterone within nuclei of the rat prostate. J. Endocr. 44, 323 (1969 b).
MAINWARING, W. I. P.: The effect of age on protein synthesis in mouse liver. Biochem. J. 113, 869 (1969 c).
MAINWARING, W. I. P.: The separation of androgen receptor and 5α-reductase activities in subcellular fractions of rat prostate. Biochem. biophys. Res. Commun. 40, 192 (1970 a).
MAINWARING, W. I. P.: Androgen receptors. Some aspects of the aetiology and biochemistry of prostatic cancer, p. 109, K. Griffiths and C. G. Pierrepoint (eds.). Cardiff: Alpha Press 1970 b.
MAINWARING, W.I.P.: A review of the formation and binding of 5α-dihydrotestosterone in the mechanism of action of androgens in the prostate of the rat and other species. J. Reprod. Fertil. 44, 377 (1975).
MAINWARING, W. I. P., BRANDES, D.: Functional and structural changes in accessory sex organs during ageing. Male accessory sex organs, p. 469, D. Brandes (ed.). New York: Academic Press 1974.
MAINWARING, W. I. P., IRVING, R. A.: The use of deoxyribose-nucleic acid-cellulose chromatography and isoelectric focusing for the characterization and partial purification of steroid-receptor complexes. Biochem. J. 134, 113 (1973).
MAINWARING, W. I. P., JONES, D. A.: Influence of receptor complexes on the properties of prostate chromatin, including its transcription by RNA polymerase. J. Steroid Biochem. 6, 475 (1975).
MAINWARING, W. I. P., MANGAN, F. R.: The specific binding of steroid-receptor complexes to DNA: evidence from androgen receptors in rat prostate. Adv. Biosci. 7, 165 (1971).
MAINWARING, W. I. P., MANGAN, F. R.: A study of the androgen receptors in a variety of androgen-sensitive tissues. J. Endocr. 59, 121 (1973).
MAINWARING, W. I. P., MANGAN, F. R., FEHERTY, P. A., FREIFELD, M.: An investigation into the antiandrogenic properties of the non-steroidal compound SCH 13521 (4′-nitro-3′-trifluoromethylisobuxryllanilide). Molec. cell. Endocr. 1, 113 (1974 a).
MAINWARING, W. I. P., MANGAN, F. R., IRVING, R. A., JONES, D. A.: Specific changes in the messenger ribonucleic acid content of the rat ventral prostate gland after androgenic stimulation: evidence from the synthesis of aldolase messenger ribonucleic acid. Biochem. J. 144, 413 (1974 b).
MAINWARING, W. I. P., MANGAN, F. R., PETERKEN, B. M.: Studies on the solubilized ribonucleic acid polymerase from rat ventral prostate gland. Biochem. J. 123, 619 (1971).
MAINWARING, W. I. P., MANGAN, F. R., WILCE, P. A., MILROY, E. J. G.: Androgens I. A review of current research on the binding and mechanism of action of androgenic steroids, notably 5α-dihydrotestosterone. Adv. exp. Med. Biol. 36, 197 (1973).
MAINWARING, W. I. P., MILROY, E. J. G.: Characterization of the specific androgen receptors in the human prostate gland. J. Endocr. 57, 371 (1973).
MAINWARING, W. I. P., PETERKEN, B. M.: A reconstituted, cell-free system for the specific transfer of steroid-receptor complexes into nuclear chromatin isolated from rat ventral prostate gland. Biochem. J. 125, 285 (1971).
MAINWARING, W. I. P., RENNIE, P. S., KEEN, J.: The androgenic regulation of prostate

proteins with a high affinity for deoxyribonucleic acid. Evidence for a prostate deoxyribonucleic acid-unwinding protein. Biochem. J. 156, 253 (1976 a).
MAINWARING, W. I. P., SYMES, E. K., HIGGINS, S. J.: Nuclear components responsible for the retention of steroid-receptor complexes, especially from the standpoint of the specificity of hormonal responses. Biochem. J. 156, 129 (1976 b).
MAINWARING, W. I. P., WILCE, P. A.: Further studies on the stimulation of protein synthesis in androgen-dependent tissues by testosterone. Biochem. J. 130, 189 (1972).
MAINWARING, W. I. P., WILCE, P. A.: The control of the form and function of the ribosomes in androgen-dependent tissues by testosterone. Biochem. J. 134, 795 (1973).
MAINWARING, W. I. P., WILCE, P. A., SMITH, A. E.: Studies on the form and synthesis of messenger ribonucleic acid in the rat ventral prostrate gland, including its tissue-specific regulation by androgens. Biochem. J. 137, 513 (1974 c).
MANGAN, F. R., MAINWARING, W. I. P.: An explanation of the antiandrogenic properties of 5α-bromo-17β-hydroxy-17α-methyl-4-oxa-5α-androstane-3-one. Steroids 20, 331 (1972).
MANGAN, F. R., MAINWARING, W. I. P.: The biochemical basis for the antagonism by BOMT of the effects of dihydrotestosterone on the rat ventral prostate gland. Cynec. Invest. 2, 300 (1973).
MANGAN, F. R., NEAL, G. E., WILLIAMS, D. C.: The effects of diethylstilboestrol and castration on the nucleic acid and protein synthesis of rat prostate gland. Biochem. J. 104, 1075 (1967).
MANGAN, F. R., NEAL, G. E., WILLIAMS, D. C.: Subcellular distribution of testosterone in rat prostate and its possible relationship to nuclear ribonucleic acid synthesis. Arch. Biochem. 124, 27 (1968).
MANGAN, F. R., PEGG, A. E., MAINWARING, W. I. P.: A reappraisal of the effects of adenosine 3′:5′-cyclic monophosphate on the function and morphology of the rat prostate gland. Biochem. J. 134, 129 (1973).
MANN, T.: The biochemistry of semen and of the male reproductive tract. New York: John Wiley and Sons 1964.
MANN, T., PARSONS, U.: Studies on the metabolism of semen. 6. Role of hormones, effect of castration, hypophysectomy and diabetes, relation between blood glucose and seminal fructose. Biochem. J. 46, 440 (1950).
MÁNYAI, S.: Isolation of the clottable protein from the secretion of the seminal vesicle of the rat. Acta physiol. Hung. 24, 419 (1964).
MÁNYAI, S., BENEY, L., CZUPPON, A.: Some characteristics of the clottable protein secreted by the seminal vesicle of the rat. Acta physiol. Acad. Sci. hung. 28, 108 (1965).
MAQUINAY, C., TIMMERMANS, L., GERBTZOFF, M. A.: Contribution à la localisation histochimique du Zn dans la prostate humaine normale et pathologique. Procès-verbaux mémoires et discussions du congrès francais d'urologie, 57, 567 (1963).
MARTIN, D. W., TOMKINS, G. M., GRANNER, D.: Synthesis and induction of tyrosine aminotransferase in synchronized hepatoma cells in culture. Proc. nat. Acad. Sci. (Wash.) 62, 248 (1969).
MARUSHIGE, K., BONNER, J.: Fractionation of liver chromatin. Proc. nat. Acad. Sci. (Wash.) 68, 2941 (1971).
MARVER, D., GOODMAN, D., EDELMAN, I. S.: Relationships between renal cytoplasmic and nuclear aldosterone receptor. Kidney Int. 1, 210 (1972).
MARYANKA, D., GOULD, H.: Transcription of rat liver chromatin with homologous enzyme. Proc. nat. Acad. Sci. (Wash.) 70, 1161 (1973).
MATHEWS, M. B., KORNER, A.: Mammalian cell-free protein synthesis directed by viral RNA. Europ. J. Biochem. 17, 328 (1970).
MATSUMOTO, K., KOTOH, K., KASAI, H., MINESITA, T., YAMAGUCHI, K.: Sub-cellular localization of radioactive steroids following administration of testosterone-^{3}H in the androgen dependent mouse tumour, Shionogi carcinoma 115. Steroids 20, 311 (1972).
MATTYSE, A. G.: Organ specificity of hormone receptor-chromatin interactions. Biochim. biophys. Acta 199, 519 (1970).

MATTYSE, A. G., ABRAMS, M.: A factor mediating interaction of kinins with the genetic material. Biochim. biophys. Acta **199**, 511 (1970).
MAWSON, C. A., FISHER, M. I.: The occurrence of zinc in the human prostate gland. Canad. J. med. Sci. **30**, 336 (1952).
MAYFIELD, J. E., BONNER, J.: Tissue differences in rat chromosomal RNA. Proc. nat. Acad. Sci. (Wash.) **68**, 2652 (1971).
MAYOL, R. F., THEYER, S. A.: Synthesis of estrogen-specific proteins in the uterus of immature rats. Biochemistry **9**, 2484 (1970).
MCCANN, S., GORLICH, L., JANSSEN, V., JUNGBLUT, P. W.: A receptor specific for 5α-dihydrotestosterone in calf uterus. Excerpta Med. Int. Cong. Ser. **210**, 150 (1970).
MCDONALD, P. G., DOUGHTY, C.: Androgen sterilization in the neonatal female rat and its inhibition by an estrogen antagonist. Neuroendocrinology **13**, 182 (1973).
MCDONALD, P., TAN, H. S., BEYER, C., SAMPSON, C., NEWTON, F., KITCHING, P., BRESCI, B., GREENHILL, R., BAKER, R., PRITCHARD, D.: Failure of 5α-dihydrotestosterone to initiate sexual behaviour in the castrated male rat. Nature (Lond.) **227**, 964 (1970).
MCGUIRE, J. S., HOLLIS, V. W., TOMKINS, G. M.: Some characteristics of the microsomal steroid reductases (5α) of rat liver. J. biol. Chem. **235**, 3112 (1960).
MCGUIRE, J. S., TOMKINS, G. M.: The heterogeneity of Δ^4-3-ketosteroid reductases (5α). J. biol. Chem. **235**, 1634 (1960).
MCGUIRE, W. L.: Estrogen receptors in human breast cancer. J. clin. Invest. **52**, 73 (1973).
MEAKIN, J. W., ROBINS, E. C.: Proceedings Third International Congress of Endocrinology, Mexico City (1968).
MENDELSON, C., DUFAU, M., CATT, K.: Gonadotrophin binding and the stimulation of cyclic adenosine 3′:5′ monophosphate and testosterone production in isolated Leydig cells. J. biol. Chem. **250**, 8818 (1975).
MERCIER-BODARD, C., ALFSEN, A., BAULIEU, E-E.: Sex steroid-binding plasma protein (SBP). Acta endocr. (Kbh.) **147**, 204 (1970).
MERCIER-BODARD, C., BAULIEU, E-E.: Récentes études des protéines plasmatiques liant la testostérone. Acta endocr. (Paris) **29**, 159 (1968).
MESTER, J., BAULIEU, E-E.: Nuclear estrogen receptor of chick liver. Biochim. biophys. Acta **261**, 236 (1972).
MICHEL, M. G., BAULIEU, E-E: Récepteur cytosoluble des androgènes dans un muscle strié squelettique. C.R. Acad. Sci. (Paris) **279**, 421 (1974).
MICKELSON, K. E., PÉTRA, P. H.: Purification of the sex steroid binding protein from human serum. Biochemistry **14**, 957 (1975).
MILGROM, E., ATGER, M., BAULIEU, E-E.: L'entré des oestrogènes dans les cellules uterines: dépend-elle d'une proteine? C.R. Acad. Sci. (Paris) **274**, 2771 (1972).
MILIN, B., ROY, A. K.: Androgen "receptor" in rat liver: cytosol "receptor" deficiency in pseudohermaphrodite male rats. Nature (Lond.) [New Biol.] **242**, 248 (1973).
MILLER, W. R., FORREST, A. P. M., HAMILTON, T. H.: Steroid metabolism by human breast and rat mammary carcinomata. Steroids **23**, 379 (1974).
MINESETA, T., YAMAGUCHI, K.: An androgen dependent mouse mammary tumour. Cancer Res. **25**, 1168 (1965).
MINGUELL, J., VALLADARES, L.: Molecular aspects on the mechanism of action of testosterone in rat bone marrow cells. J. Steroid Biochem. **5**, 649 (1974).
MIRSKY, A. E., OSAWA, S.: The interphase nucleus. The cell, vol. 2, p. 677, A. E. Mirsky and J. Brachet (eds.). New York: Academic Press 1961.
MOBBS, B. G., JOHNSON, I. E., CONNOLLY, J. G.: Hormonal responses of prostatic carcinoma. Urology **3**, 105 (1974).
MOBBS, B. G., JOHNSON, I. E., CONNOLLY, J. G.: *In vitro* assay of androgen binding by human prostate. J. Steroid Biochem. **6**, 453 (1975).
MOHLA, S., DE SOMBRE, E. R., JENSEN, E. V.: Tissue-specific stimulation of RNA synthesis by transformed estradiol-receptor complex. Biochem. biophys. Res. Commun. **46**, 661 (1972).

MOLINEUX, I. J., GEFTER, M. L.: Properties of the *E. coli* DNA binding (unwinding) protein: interaction with DNA polymerases and DNA. Proc. nat. Acad. Sci. (Wash.) 71, 3858 (1974).

MOLLOY, G. R., THOMAS, W. L., DARNELL, J. E.: Occurrence of uridylate-rich oligonucleotide regions in heterogeneous nuclear RNA of HeLa cells. Proc. nat. Acad. Sci. (Wash.) 69, 3684 (1972).

DE MOOR, P., DENEF, C.: The "puberty" of the rat liver. Female pattern of cortisol metabolism in male rats castrated at birth. Endocrinology 82, 480 (1968).

MOORE, C. R., HUGHES, W., GALLAGHER, T. F.: Rat seminal vesicle cytology as a testis-hormone indicator and the prevention of castration changes by testis extract injection. Amer. J. Anat. 45, 109 (1930).

MOORE, R. J., GRIFFIN, J. E., WILSON, J. D.: Diminished 5α-reductase activity in extracts of fibroblasts cultured from patients with familial incomplete male pseudohermaphroditism, type 2. J. biol. Chem. 250, 7168 (1975).

MOORE, R. J., WILSON, J. D.: The effect of androgenic hormones on the reduced nicotinamide adenine dinucleotide phosphate: Δ^4-3-ketosteroid 5α-oxidoreductase of rat ventral prostate. Endocrinology 93, 581 (1973 a).

MOORE, R. J., WILSON, J. D.: Localization of the reduced nicotinamide adenine dinucleotide phosphate: Δ^4-3-ketosteroid 5α-oxidoreductase in the nuclear membrane of the rat ventral prostate. J. biol. Chem. 247, 958 (1973 b).

MORFIN, R. F., ALIAPOULIOS, M. A., CHAMBERLAIN, J., OFNER, P.: Metabolism of testosterone-4-^{14}C by the canine prostate and urinary bladder *in vivo*. Endocrinology 87, 394 (1970).

MORGAN, M. D., WILSON, J. D.: Intranuclear metabolism of progesterone-1,2-^{3}H in the hen oviduct. J. biol. Chem. 245, 3781 (1970).

MORHENN, V., RABINOVITZ, Z., TOMKINS, G. M.: Effects of adrenal glucocorticoids on polyoma virus replication. Proc. nat. Acad. Sci. (Wash.) 70, 1088 (1973).

MORISHITA, H., NAFTOLIN, F., TODD, R. B., WILEN, R., DAVIES, I. J., RYAN, K. J.: Lack of an effect of dihydrotestosterone on serum luteinizing hormone in neonatal female rats. J. Endocr. 67, 139 (1975).

MORLEY, A. R., WRIGHT, N. A., APPLETON, D.: Cell proliferation in the castrate mouse seminal vesicle in response to testosterone propionate. 1. Experimental observations. Cell Tiss. Kinet. 6, 239 (1973).

MORRISON, R. L., JOHNSON, D. C.: The effects of androgenization in male rats castrated at birth. J. Endocr. 34, 117 (1966).

MOWSZOWICZ, I., BARDIN, C. W.: *In vivo* androgen metabolism in mouse kidney. High 3-keto-reductase (3α-hydroxysteroid dehydrogenase) activity relative to 5α-reductase. Steroids 23, 793 (1974).

MULDER, E., BRINKMANN, A. O., LAMERS, STAHLHOFEN, G. J. M., VAN DER MOLEN, H. J.: Binding of estradiol by the nuclear fraction of rat interstitial tissue. FEBS Lett. 31, 131 (1973).

MULDER, E., PETERS, M. J., DE VRIES, J., VAN DER MOLEN, H. J.: Characterization of a nuclear receptor for testosterone in seminiferous tubules of mature rat testis. Molec. cell. Endocr. 2, 171 (1975).

MULDOON, T. G., WESTPHAL, U.: Steroid-protein interactions. XV. Isolation and characterization of corticosteroid-binding globulin from human plasma. J. biol. Chem. 242, 5636 (1967).

MUNCK, A., BRINCK-JOHNSEN, T.: Specific and non-specific physicochemical interactions of glucocorticoids and related steroids with rat thymus cells *in vitro*. J. biol. Chem. 243, 5556 (1968).

MUNCK, A., BRINCK-JOHNSEN, T.: Is cortisol metabolized as it dissociates from glucucorticoid receptors in thymus cells. J. Steroid Biochem. 5, 203 (1974).

MUNCK, A., WIRA, C.: Glucocorticoid receptors in rat thymus cells. Advanc. Biosci. 7, 301 (1971).

MÜNTZING, J., SHUKLA, S. K., CHU, T. M., MITTLEMAN, A., MURPHY, G. P.: Pharmacological study of oral estramustine phosphate (Estracyt) in advanced carcinoma of the prostate. Invest. Urol. 12, 65 (1974).

MURAMATSU, M., SHIMADA, N., HIGASHINAKAGAWA, T.: Effect of cycloheximide on the nucleolar RNA synthesis in rat liver. J. molec. Biol. 53, 91 (1970).

MURPHY, B. E. P.: Binding of testosterone and estradiol in plasma. Canad. J. Biochem. 46, 299 (1968).

NAFTOLIN, F., FEDER, H. H.: Suppression of luteinizing hormone secretion in male rats by 5α-androstan-17β-ol-3-one (dihydrotestosterone) propionate. J. Endocr. 56, 155 (1973).

NAFTOLIN, F., RYAN, K. J., PETRO, Z.: Aromatization of androstenedione by the limbic system tissue from human foetuses. J. Endocr. 51, 795 (1971).

NEAL, G. E.: Androgen uptake by rat ventral prostate. Biochem. J. 118, 12P.

NEEDHAM, J., LU, G. D.: Sex hormones in the middle ages. Endeavour, 27, 130 (1968).

NEMER, M., DUBROFF, L. M., GRAHAM, M.: Properties of sea urchin embryo messenger RNA containing and lacking poly(A). Cell 6, 171 (1975).

NERI, R., FLORANCE, K., KOZIOL, P., VAN CLEAVE, S.: A biological profile of a non-steroidal antiandrogen SCH 13521 (4′-nitro-3′-trifluoroisobutyrlanilide). Endocrinology 91, 427 (1972).

NERI, R., MONAHAN, M.: Effects of a novel nonsteroidal antiandrogen on canine prostatic hyperplasia. Invest. Urol. 10, 123 (1972).

NICANDER, L.: An electron microscopical study of absorbing cells in the posterior caput epididymidis of rabbits. Z. Zellforsch. 66, 829 (1965).

NIEMI, M., TUOHIMAA, P.: The mitogenic activity of testosterone in the accessory sex glands of the rat, in relation to its conversion to testosterone. Basic actions of sex steroids on target organs, p. 258, P. O. Hubinont, C. Leroy and P. Galand (eds.). Basel: S. Karger 1971.

NIMROD, A., RYAN, K. J.: Aromatization of androgens by human abdominal and breast fat tissue. J. clin. Endocr. 40, 367 (1975).

NOLL, M.: Subunit structure of chromatin. Nature (Lond.) 251, 249 (1974).

NOMA, K., SATO, B., YANO, S., YAMAMURA, Y., SEKI, T.: Metabolism of testosterone in the hypothalamus of the male rat. J. Steroid Biochem. 6, 1261 (1975).

NORDENSKJÖLD, B. A., SKOOG, L., BROWN, N. C., REICHARD, P.: Deoxyribonucleotide pools and deoxyribonucleic acid synthesis in cultured mouse embryo cells. J. biol. Chem. 245, 5360 (1970).

NORTHCUTT, R. C., ISLAND, D. P., LIDDLE, G. W.: An explanation of the target organ insensitiveness to testosterone in the testicular feminization syndrome. J. clin. Endocr. 29, 422 (1969).

NOTIDES, A., GORSKI, J.: Estrogen-induced synthesis of a specific uterine protein. Proc. nat. Acad. Sci. (Wash.) 56, 230 (1966).

NOTIDES, A., WILLIAMS-ASHMAN, H. G.: The basic protein responsible for the clotting of guinea pig semen. Proc. nat. Acad. Sci. (Wash.) 58, 1991 (1967).

NOUMURA, T., WEISZ, J., LLOYD, C. W.: *In vitro* conversion of 7-^{3}H-progesterone to androgens by the rat testis during the second half of fetal life. Endocrinology 78, 245 (1966).

NOZU, K., TAMAOKI, B-I.: Incorporation of ^{131}I-labelled androgen receptor into nuclei of rat prostate. Biochem. biophys. Res. Commun. 58, 145 (1974).

NOZU, K., TAMAOKI, B-I.: On the role of the cytosol receptor in the incorporation of androgens into the prostatic nuclei of the rat. J. Steroid Biochem. 6, 57 (1975 a).

NOZU, K., TAMAOKI, B-I.: Formation, nuclear incorporation and enzymatic decomposition of the androgen receptor of rat prostate. J. Steroid Biochem. 6, 1319 (1975 b).

NUNEZ, E., SAVU, L., ENGLEMANN, L., BENASSAYAG, C., CRÉPY, O., JAYLE, M. F.: Origine embryonaire de la protéine sérique fixant l'oestrone et l'oestradiol chez la ratte impubère. C.R. Acad. Sci. (D) Paris 273, 242 (1971).

OHNO, S., DOFUKU, R., TETTENBORN, U.: More about the x-linked testicular feminization of the mouse as a noninducible (i^{s}) mutation of a regulatory locus: 5α-androstan-3α,17β-diol as the true inducer of kidney alcohol dehydrogenase and β-glucuronidase. Clin. Genet. 2, 128 (1971).

OHNO, S., LYON, M. F.: X-linked testicular feminization in the mouse as a noninducible regulatory mutation of the Jacob-Monod type. Clin. Genet. 1, 121 (1970).

OHNO, S., STENIUS, S., CHRISTIAN, L., HARRIS, C., IVEY, C.: More about the testosterone induction of kidney alcohol dehydrogenase in the mouse. Biochem. Genet. 4, 565 (1970).

O'FARELL, P. H., DANIEL, J. C.: Estrogen binding in the uteri of mammals with delayed implanation. Endocrinology 88, 1104 (1971).

OFNER, P.: Effects and metabolism of hormones in normal and neoplastic prostate tissue. Vitam. and Horm. 26, 237 (1968).

OLINS, A. L., OLINS, D. E.: Spheroid chromatin units (ν bodies). Science 183, 330 (1974).

ORLANDINI, G.: Effects of castration on mitochondria structure in accessory glands of the mouse. Arch. ital. Anat. Embriol. 71, 149 (1964).

OSHIMA, H., SARADA, T., OCHIAI, K., TAMAOKI, B-I.: Δ^4-5α-hydrogenase in immature rat testes: its intracellular distribution, enzyme kinetics and influence of administered gonadotrophin and testosterone propionate. Endocrinology 86, 1215 (1970).

OSHIMA, H., WAKABAYASHI, K., TAMAOKI, B-I.: The effect of synthetic estrogen upon the biosynthesis *in vitro* of androgen and luteinizing hormone in the rat. Biochim. biophys. Acta 137, 356 (1967).

OUDET, P., GROSS-BELLARD, M., CHAMBON, P.: Electron microscopic and biochemical evidence that chromatin structure is a repeating unit. Cell 4, 281 (1975).

PAIGEN, K.: The effect of mutation on the intracellular location of β-glucuronidase. Exp. Cell Res. 25, 286 (1961 a).

PAIGEN, K.: The genetic control of enzyme activity during differentiation. Proc. nat. Acad. Sci. (Wash.) 47, 1641 (1961 b).

PAIGEN, K., SWANK, R. T., TOMINO, S., GANSCHOW, R. E.: The molecular genetics of mammalian glucuronidase. J. cell. Physiol. 85, 379 (1975).

PÂRÍZEK, J.: The destructive effect of cadmium on testicular tissue and its prevention by zinc. J. Endocr. 15, 56 (1957).

PARSONS, I. C.: The metabolism of testosterone by early chick embryonic blastoderm. Steroids 16, 59 (1970).

PASQUALINI, J. R., SUMIDA, L., GELLY, C.: Mineralocorticoid receptors in the foetal compartment. J. Steroid Biochem. 3, 543 (1972).

PAUL, J.: DNA marking in mammalian chromatin: a molecular mechanism for determination of cell type. Curr. Topics Develop. Biol. 5, 317 (1970).

PAUP, D. C., CONIGLIO, L. P., CLEMENS, L. G.: Masculinization of the female golden hamster by neonatal treatment with androgen or estrogen. Horm. Behav. 3, 123 (1972).

PEETS, E. A., HENSEN, M. F., NERI, R.: On the mechanism of the anti-androgenic action of flutamide (α,α,α-trifluoro-2-methyl-4'-nitro-m-propiotoluidide) in the rat. Endocrinology 94, 532 (1974).

PEGG, A. E., LOCKWOOD, D. H., WILLIAMS-ASHMAN, H. G.: Concentrations of putrescine and polyamines and their enzymic synthesis during androgen-induced prostatic growth. Biochem. J. 117, 17 (1970).

PEGG, A. E., WILLIAMS-ASHMAN, H. G.: Effects of androgens on incorporation of labelled amino acids into proteins by prostate mitochondria. Endrocrinology 82, 603 (1968).

PEGG, A. E., WILLIAMS-ASHMAN, H. G.: On the role of S-adenosyl-L-methionine in the biosynthesis of spermidine by rat prostate. J. biol. Chem. 244, 682 (1969).

PENIT, C., PARAF, A., CHAPEVILLE, F.: Terminal deoxynucleotidyl transferase in murine plasmacytomas. Nature (Lond.) 249, 755 (1974).

PENNEQUIN, P., ROBEL, P., BAULIEU, E-E.: Steroid-induced early protein synthesis in rat uterus and prostate. Eur. J. Biochem. 60, 137 (1975).

PÉREZ-PALACIOS, G., MORATO, T., PÉREZ, A. E., CASTEÑADA, E., GUAL, G.: Biochemical studies on the incomplete form of testicular feminization syndrome. Steroids 17, 471 (1971).

PÉZARD, A.: Biologie générale sur la determination des characteurs sexuels secondaire chez les gallinacés. C. R. Acad. Sci. (Paris) **153**, 1027 (1911).

PIKE, A., PELLING, W. B., HARPER, M. E., PIERREPOINT, C. G., GRIFFITHS, K.: Testosterone metabolism *in vivo* by human prostatic tissue. Biochem. J. **120**, 443 (1970).

PLAPINGER, L., MCEWEN, B. S.: Ontogeny of estradiol-binding sites in rat brain. I. Appearance of presumptive adult receptors in cytosol and nuclei. Endocrinology **93**, 1119 (1973).

POHLMAN, C., HALLBERG, M. C., ZORN, E. M., GUEVARA, A., WIELAND, R. G.: Testosterone binding affinity: alterations caused by pathological and physiological states. Amer. J. med. Sci. **258**, 270 (1969).

POLLARD, M., LUCKERT, P. H.: Transplantable metastasizing prostate adenocarcinomas in rats. J. nat. Cancer Inst. **54**, 643 (1975).

POORTMAN, J., PRENEN, J. A. C., SCHWARTZ, F., THIJSSEN, J. H. H.: Interaction of Δ^5-androstene-$3\beta,17\beta$-diol with estradiol and dihydrotestosterone in human myometrial and mammary cancer tissue. J. clin. Endocr. **40**, 373 (1975).

PORTER, A., CAREY, N., FELLNER, P.: Presence of a large poly(rC) tract in the RNA of encephalomyocarditis virus. Nature (Lond.) **248**, 675 (1974).

POWERS, M. L., FLORINI, J. R.: A direct effect of testosterone on muscle cells in tissue culture. Endocrinology **97**, 1043 (1975).

DE PRAW, E. J.: DNA and chromosomes. New York: Holt, Rinehart, Winston 1970.

PRICE, D., PANNABECKER, R.: Comparative responsiveness of homologous sex ducts and accessory glands of fetal rats in culture. Arch. Anat. micros. Morphol. Exp. **48**, 223 (1959).

PRICE, D., WILLIAMS-ASHMAN, H. G.: The accessory reproductive glands of mammals. Sex and internal secretions, 3rd edition, vol. I, p. 366, W. C. Young (ed.). Baltimore: Williams and Wilkins Co., 1961.

PUCA, G. A., NOLA, E., CASO, P., SICA, V.: Interaction of estrogen receptor with chromatin. Proceedings XI International Cancer Congress Florence, vol. 5, p. 358, P. Buccalosi, U. Veronesi, N. Cascinelli (eds.). Amsterdam: Excerpta Medica 1975 a.

PUCA, G. A., NOLA, E., HIBNER, U., CICALA, G., SICA, V.: Interactions of the estradiol complex from calf uterus with its nuclear acceptor sites. J. biol. Chem. **250**, 6542 (1975 b).

PUCA, G. A., NOLA, E., SICA, V., BRESCIANI, F.: Estrogen-binding proteins of calf uterus. Interrelationship between various forms and identification of a receptor-transforming factor. Biochemistry **11**, 4157 (1972).

PUCA, G. A., SICA, V., NOLA, E.: Identification of a high-affinity nuclear acceptor site for estrogen receptors of calf uterus. Proc. nat. Acad. Sci. (Wash.) **71**, 979 (1974).

PULLEYBANK, D. E., MORGAN, A. R.: Partial purification of "ω" protein from calf thymus. Biochemistry **14**, 5205 (1975).

RAMM, E. I., VOROBEV, V. I., BIRSHSTEIN, T. M., BOLOTINA, I. A., VOLKENSCHTEIN, M. V.: Circular dichroism of DNA and histone in the free state and in deoxyribonucleoprotein. Europ. J. Biochem. **25**, 245 (1972).

RAYNAUD, J. P., MERCIER-BODARD, C., BAULIEU, E-E.: Rat estradiol binding plasma protein (EBP). Steroids **18**, 767 (1971).

REDDY, V. V. R., NAFTOLIN, F., RYAN, K. J.: Conversion of androstenedione to estrone by neural tissues from fetal and neonatal rats. Endrocrinology **94**, 117 (1974).

REECK, G. R., SIMPSON, R. T., SOBER, H. A.: Resolution of a spectrum of nucleoprotein species in sonicated chromatin. Proc. nat. Acad. Sci. (Wash.) **69**, 2317 (1972).

REED, M. J., STITCH, S. R.: The uptake of testosterone and zinc *in vitro* by the human benign hypertrophic tissue. J. Endocr. **58**, 405 (1973).

REEDER, R. R.: Transcription of chromatin by bacterial RNA polymerase. J. molec. Biol. **80**, 229 (1973).

REEL, J. R., VAN DEWARCK, S. D., SHI, Y., CALLATINE, M. R.: Macromolecular binding and metabolism of progesterone in the decidual and pseudopregnant rat and rabbit uterus. Steroids **18**, 441 (1971).

RENNIE, P. S., BRUCHOVSKY, N.: *In vitro* and *in vivo* studies on the functional significance of androgen receptors in rat prostate. J. biol. Chem. **247**, 1546 (1972).

RENNIE, P. S., SYMES, E. K., MAINWARING, W. I. P.: The androgenic regulation of the activities of enzymes engaged in the synthesis of deoxyribonucleic acid in rat ventral prostate gland. Biochem. J. **152**, 1 (1975).

RENZ, M.: Preferential and cooperative binding of histone I to chromosomal mammalian DNA. Proc. nat. Acad. Sci. (Wash.) **72**, 733 (1975).

RESNICK, M. I., WALVOORD, D. J., GRAYHACK, J. T.: Effect of prolactin on testosterone uptake by the perfused canine prostate. Surg. Forum **25**, 70 (1974).

RICHARDS, B. M., PARDON, J. F.: The molecular structure of nucleohistone (DNH). Exp. Cell Res. **62**, 184 (1970).

RICHTER, K. H., SEKERIS, C. E.: Isolation and partial purification of non-histone chromosomal proteins from rat liver, thymus and kidney. Arch. Biochem. **148**, 44 (1972).

RIGGS, A. D., BOURGEOIS, S., COHN, M.: The lac repressor-operator interactions III. Kinetic studies. J. molec. Biol. **53**, 401 (1970).

RILL, R., VAN HOLDE, K. E.: Properties of nuclease-resistant fragments of calf thymus chromatin. J. biol. Chem. **248**, 1080 (1973).

RITZÉN, E. M., DOBBINS, E. C., TINDALL, D. J., FRENCH, F. S., NAYFEH, S. N.: Characterization of an androgen-binding protein (ABP) in rat testis and epididymis. Steroids **21**, 593 (1973).

RITZÉN, E. M., NAYFEH, S. N., FRENCH, F. S., DOBBINS, M. C.: Demonstration of androgen-binding components in rat epididymis cytosol and comparison with binding components in prostate and other tissues. Endocrinolgy **89**, 143 (1971).

RITZÉN, E. M., SHIHADEEN, N. N., FRENCH, F. S., ARONIN, P. A.: Deficient nuclear uptake of testosterone in the androgen insensitive (Stanley-Gumbreck) pseudohermaphrodite male rat. Endocrinology **91**, 116 (1972).

ROARK, D. E., GEOGHEGAN, T. E., KELLER, G. H.: A two-subunit histone complex from calf thymus. Biochem. biophys. Res. Commun. **59**, 542 (1974).

ROBBINS, E., BORUN, T. W.: The cytoplasmic synthesis of histones in HeLa cells and its temporal relationship to DNA synthesis. Proc. nat. Acad. Sci. (Wash.) **57**, 409 (1967).

ROBEL, P., BLONDEAU, J-P., BAULIEU, E-E.: Androgen receptors in rat ventral prostate microsomes. Biochim. biophys. Acta **373**, 1 (1974).

ROBEL, P., CORPECHOT, C., BAULIEU, E-E.: Testosterone and androstanolone in rat plasma and tissues. FEBS Lett. **33**, 218 (1973).

ROBERTS, J.: Termination factor for RNA synthesis. Nature (Lond.) **224**, 1168 (1969).

ROBINSON, M. R. G., THOMAS, B. S.: Effect of hormonal therapy on plasma testosterone levels in prostatic carcinoma. Brit. med. J. **4**, 391 (1971).

ROCHEFORT, H.: Aspects of estrogen receptors in rat uterus. Excerpta Medica Int. Congr. Ser. **219**, 376 (1971).

ROMANOV, A. L.: The avian embryo. New York: Macmillan Co. 1960.

ROMMERTS, F. F. G., VAN DER MOLEN, H. J.: Occurrence and localization of 5α-steroid reductase, 3α- and 17β-hydroxysteroid dehydrogenases in hypothalamus and other brain tissues of the male rat. Biochim. biophys. Acta **248**, 489 (1971).

ROSEN, V., JUNG, I., BAULIEU, E-E., ROBEL, P.: Androgen binding protein in human benign prostatic hypertrophy. J. clin. Endocr. **41**, 761 (1975).

ROSNER, W., SMITH, R. N.: Isolation and characterization of the testosterone-oestradiol-binding globulin from human plasma. Use of a novel affinity column. Biochemistry **14**, 4813 (1975).

ROUSSEAU, G. G., BAXTER, J. D., TOMKINS, G. M.: Glucocorticoid receptors: relations between steroid binding and biological effect. J. molec. Biol. **67**, 99 (1972).

ROY, A. K., BAULIEU, E-E., FEYEL-CABANES, T., LE GOASCOIGNE, C., ROBEL, P.: Hormone metabolism and action. II. Androstenedione in prostate organ culture. Endocrinology **91**, 396 (1972 a).

ROY, A. K., ROBEL, P., BAULIEU, E-E.: 3α-Hydroxy and 3β-hydroxy C_{19} steroids in prostate organ culture. Endocrinology **92**, 1216 (1972 b).

RUH, T. S., RUH, M. F.: Androgen induction of a specific uterine protein. Endocrinology 97, 1144 (1975).

RUSSELL, D. H.: Putrescine and spermidine biosynthesis in the development of normal and anucleolate mutants of Xenopus laevis. Proc. nat. Acad. Sci. (Wash.) 68, 523 (1971).

RUSSELL, D. H.: Polyamines in normal and neoplastic growth. New York: Raven Press 1973.

RUTTER, W. J.: Electrophoresis of aldolases. Fed. Proc. 23, 1248 (1964).

SALAS, J., GREEN, H.: Proteins binding to DNA and their relation to growth in cultured mammalian cells. Nature (Lond.) [New Biol.] 229, 165 (1971).

SAMPAREZ, S., THIEULANT, M. L., MERCIER, L., JUOAN, P.: A specific testosterone receptor in the cytosol of rat anterior hypophysis. J. Steroid Biochem. 5, 911 (1974).

SAMUELS, H. H., TOMKINS, G. M.: Relation of steroid structure to enzyme induction in hepatoma tissue culture cells. J. molec. Biol. 52, 57 (1970).

SAMUELS, L. T., HARDING, B. W., MANN, T.: Aldose reductase and ketose reductase in male accessory organs of reproduction. Distribution and relation to seminal fructose. Biochem. J. 84, 39 (1962).

SAR, M., LIAO, S., STUMPF, W. E.: Nuclear concentration of androgens in rat seminal vesicles and prostate demonstrated by dry-mount autoradiography. Endocrinology 86, 1008 (1970).

SAR, M., STUMPF, W. E.: Autoradiographic localization of radioactivity in the rat brain after injection of 1,2-^{3}H-testosterone. Endocrinolgy 92, 251 (1973 a).

SAR, M., STUMPF, W. E.: Cellular and subcellular localization of radioactivity in the rat pituitary after injection of 1,2-^{3}H-testosterone using dry-mount autoradiography. Endocrinology 92, 631 (1973 b).

SAUNDERS, F. J.: Some aspects of relation of structure of steroids to their prostate-stimulating effects. Nat. Cancer Inst. Monogr. 12, 139 (1963).

SAUNDERS, F. J.: Effects of sex steroids and related compounds on pregnancy and on development of the young. Physiol. Rev. 48, 601 (1968).

SAVLOV, E. D., WITTLIFF, J. L., HILF, R., HALL, T. C.: Correlations between certain biochemical properties of breast cancer and response to therapy: a preliminary report. Cancer (Philad.) 33, 303 (1974).

SCATCHARD, G.: The attractions of proteins for small molecules and ions. Ann. N.Y. Acad. Sci. 51, 660 (1949).

SCHACHMAN, H. K.: Ultracentrifugation in biochemistry. New York: Academic Press 1959.

SCHINDLER, A. E., EBERT, A., FREIDRICH, E.: Conversion of androstenedione to estrone by human fat tissue. J. clin. Endocr. 35, 627 (1972).

SCHOCHETMAN, G., PERRY, R. P.: Early appearance of histone messenger RNA in polyribosomes of cultured L-cells. J. molec. Biol. 63, 591 (1972).

SCHWARTZ, R. J., TSAI, M-J., TSAI, S. Y., O'MALLEY, B. W.: Effect of estrogen on gene expression in chick oviduct. V. Changes in the number of RNA polymerase binding and initiation sites in chromatin. J. biol. Chem. 250, 5175 (1975).

SCHWEIKERT, H. U., MILEWICH, L., WILSON, J. D.: Aromatization of androstenedione by human hairs. J. clin. Endocr. 40, 413 (1975).

SCOW, R. O., HAGEN, S. N.: Effect of testosterone propionate and growth hormone on growth and chemical composition on muscle and other tissues in hypophysectomized male rats. Endocrinology 77, 852 (1965).

SEGAL, S. J., DAVIDSON, O. W., WADA, K.: Role of RNA in the regulatory action of estrogen. Proc. nat. Acad. Sci. (Wash.) 54, 782 (1965).

SEGALOFF, A.: The enhanced local androgenic activity of 19-norsteroids and stabilization of their structure by 7α- and 17α-methyl substituents. Steroids 1, 299 (1963).

SEPSENWOL, S., HECHTER, O.: Failure to observe testosterone-induced nucleus-lysosome interactions in rat ventral prostate. Molec. cell. Endocr. 4, 115 (1976).

SHAIN, S. A., AXELROD, L. R.: Reduced high-affinity 5α-dihydrotestosterone receptor capacity in the ventral prostate of the ageing rat. Steroids 21, 801 (1973).

SHAO, T-C., CASTENADA, E., ROSEFIELD, R. L., LIAO, S.: Selective retention and formation of a Δ^5-androstenediol-receptor complex in cell nuclei of the rat vagina. J. biol. Chem. **250,** 3095 (1975).

SHARP, P. A., SUGDEN, B., SAMBROOK, J.: Determination of two restriction endonuclease activities in *Haemophilus parainfluenzae* using analytical agarose-ethidium bromide electrophoresis. Biochemistry **12,** 3055 (1973).

SHELDON, R., JURALE, C., KATES, J.: Detection of polyadenylic acid sequences in viral and eukaryotic RNA. Proc. nat. Acad. Sci. (Wash.) **69,** 417 (1972).

SHIMAZAKI, J., KURIHARA, H., ITO, Y., SHIDA, K.: Testosterone metabolism in prostate: formation of androstan-17β-ol-3-one and androst-4-en-3,17-dione and inhibiting effect of natural and synthetic oestrogen. Gunma J. med. Sci. **14,** 313 (1965).

SHIMAZAKI, J., MATSUSHITA, I., FURUYA, N., YAMAKA, H., SHIDA, K.: Reduction of 5α-position of testosterone in the rat ventral prostate. Endocr. jap. **16,** 453 (1969).

SHIMAZAKI, J., SATO, J., NAGAI, H., SHIDA, K.: Effects of inhibitors of nucleic acid and protein biosynthesis on the rate of 5α-reduction of testosterone: the activity of DNA polymerase and nucleic acid contents of testosterone-stimulated prostate of rats. Endocr. jap. **17,** 175 (1970).

SHIMAZAKI, J., TAGUCHI, I., YAMANAKA, H., MAYUZUMI, T., SHIDA, K.: Effects of testosterone administration on the free amino acids and conversion of arginine to proline and glutamate in the rat ventral prostate. Endocr. jap. **20,** 455 1973).

SHOLITON, L. J., JONES, C. L., WERK, E. E.: The uptake and metabolism of [1,2-^{3}H] testosterone by the brain of functionally hepatectomized and totally eviscerated male rats. Steroids **20,** 399 (1972).

SICA, V., PARIKH, I., NOLA, E., PUCA, G. A., CUATRECASAS, P.: Affinity chromatography and the purification on estrogen receptors. J. biol. Chem. **248,** 6543 (1973).

SIERRALTA, W., GONSALEZ, M. C., MINGUELL, J.: The effect of testosterone on bone marrow nuclear ribonucleic acid metabolism. J. Steroid Biochem. **5,** 645 (1974).

SIGAL, N., DELIUS, H., KORNBERG, T., GEFTER, M. L., ALBERTS, B. M.: A DNA-unwinding protein isolated from *Escherichia coli:* its interaction with DNA and with DNA polymerases. Proc. nat. Acad. Sci. (Wash.) **69,** 3537 (1972).

SINGHAL, R. L., PARVLEKAR, M. R., VIJAYVARGIYA, R., ROBISON, G. A.: Metabolic control mechanisms in mammalian systems. Biochem. J. **125,** 329 (1971).

SKINNER, R. W. S., POZDERAC, R. V., COUNSELL, R. E., WEINHOLD, P. M.: The inhibitive effects of steroid analogues in the binding of tritiated 5α-dihydrotestosterone to receptor proteins from rat prostate tissue. Steroids **25,** 185 (1975).

SMITH, A. A., MCLEAN, W. S., HANSSON, V., NAYFEH, S. N., FRENCH, F. S.: Androgen receptor in the nuclei of rat testis. Steroid **25,** 569 (1975).

SMITH, J. A., KING, R. J. B.: Effects of steroids on the growth of an androgen-dependent mouse mammary tumour in cell culture. Exp. Cell Res. **73,** 351 (1972).

SMITH, J. A., MARTIN, L.: Do cells cycle? Proc. nat. Acad. Sci. U. S. A. **70,** 1263 (1973).

SMITH, R. G., IRAMAIN, C. A., BUTTRAM, V. C., O'MALLEY, B. W.: Purification of human uterine progesterone receptor. Nature (Lond.) [New Biol.] **253,** 217 (1975).

DE SOMBRE, E. R., SMITH, S., BLOCK, G. E., FERGUSON, D. J., JENSEN, E. V.: Prediction of breast cancer response in human breast cancer. Cancer Chemother. Rep. **58,** 513 (1974).

SOUTHREN, A. L., TOCHIMOTO, S., CARMODY, N. C., ISURGI, K.: Plasma production rates of testosterone in normal adult men and women and in patients with the syndrome of feminizing testes. J. clin. Endocr. **25,** 1441 (1965).

SPATZ, L., STRITTMATTER, R.: A form of reduced nicotinamide adenine dinucleotide cytochrome b 5 reductase containing both the catalytic site and an additional hydrophobic membrane-binding segment. J. biol. Chem. **248,** 793 (1973).

SPELSBERG, T. C., STEGGLES, A. W., CHYTIL, F., O'MALLEY, B. W.: Progesterone-binding components of chick oviduct. V. Exchange of progesterone-binding capacity from target to non-target tissue chromatins. J. biol. Chem. **247,** 1368 (1972).

SPELSBERG, T. C., STEGGLES, A. W., O'MALLEY, B. W.: Progesterone-binding com-

ponents of chick oviduct. III. Chromatin acceptor sites. J. biol. Chem. **246**, 4188 (1971).
STANLEY, A. J., GUMBRECK, L. G.: A pseudohermaphrodite mutant of Oklahoma rats. Proc. Amer. endocr. Soc. **40** (1964).
STARLING, E. H.: The chemical correlation of the functions of the body. Lancet **ii**, 339 (1905).
STEELMAN, R., BROOKS, J. R., MORGAN, E. R., PATENELLI, D. J.: Anti-androgenic activity of spironolactone. Steroids **14**, 449 (1969).
STEGGLES, A. W., SPELSBERG, T. C., GLASSER, S. R., O'MALLEY, B. W.: Soluble complexes between steroid hormones and target-tissue receptors bind specifically to target-tissue chromatin. Proc. nat. Acad. Sci. (Wash.) **68**, 1482 (1971).
STEGGLES, A. W., WILSON, G. N., KANTOR, J. J., PICCIANO, D. J., FALVEY, A. K., ANDERSON, W. F.: Cell-free transcription of mammalian chromatin; transcription of globin messenger RNA sequences from bone marrow chromatin with mammalian RNA polymerase. Proc. nat. Acad. Sci. (Wash.) **71**, 1219 (1974).
STEINBERGER, E.: Hormonal control of mammalian spermatogenesis. Physiol. Rev. **51**, 1 (1971).
STEINITZ, B. G., GIANNINI, T., BUTLER, M., POPICK, F.: Dissociation of anabolic and androgenic properties of steroids by anti-anabolic and anti-androgenic agents in rats. Endocrinology **89**, 894 (1971).
STEINS, P., KREIG, M., HOLLMANN, H. J., VOIGT, K. D.: *In vitro* studies of testosterone and 5α-dihydrotestosterone binding in benign prostatic hypertrophy. Acta endocr. (Kbh.) **75**, 773 (1974).
STRITTMATTER, R., ROGERS, M. J., SPATZ, L.: The binding of cytochrome 65 to liver microsomes. J. biol. Chem. **247**, 7188 (1972).
STROUPE, S. D., WESTPHAL, U.: Steroid-protein interactions. Stopped flow fluorescence studies of the interactions between hormones and progesterone-binding globulin. J. biol. Chem. **250**, 8735 (1975).
SUFRIN, G., COFFEY, D. S.: A new model for studying the effect of drugs on prostatic growth. 1. Antiandrogens and DNA synthesis. Invest. Urol. **11**, 45 (1973).
SULLIVAN, J. N., STROTT, C. A.: Evidence for an androgen-independent mechanism regulating the levels of receptor in target tissue. J. biol. Chem. **248**, 3202 (1973).
SUTHERLAND, D. J. A., ROBINS, E. C. MEAKIN, J. W.: Effect of androgens on Shionogi 115 *in vitro*. J. nat. Cancer Inst. **52**, 37 (1974).
SWANK, R. T., BAILEY, D. W.: Recombinant inbred lines: value in the gentic analysis of biochemical variants. Science **181**, 1249 (1973).
SWANK, R. T., PAIGEN, K.: Biochemical and genetic evidence for a macromolecular β-glucuronidase complex in microsomal membranes. J. molec. Biol. **77**, 371 (1973).
SWANK, R. T., PAIGEN, K., GANSCHOW, R. E.: Genetic control of glucuronidase induction in mice. J. molec. Biol. **81**, 225 (1973).
SWERDLOFF, R. S., WALSH, P. C., O'DELL, W. D.: Control of LH and FSH secretion in the male: evidence that aromatization of androgens to oestradiol is not required for inhibition of gonadotrophin secretion. Steroids **20**, 13 (1972).
SZEGO, C. M.: The lysosome as a mediator of hormone action. Recent Progr. Horm. Res. **30**, 171 (1974).
SZEGO, C. M., DAVIS, J. S.: Adenosine 3′, 5′-monophosphate in rat uterus: acute elevation by estrogen. Proc. nat. Acad. Sci. (Wash.) **58**, 1711 (1967).
TABEI, T., HEINRICHS, W. L.: Testosterone as a neonatal determinant in rats of the 7- and 16α-hydroxylases and reduction of 3β-hydroxyandrost-5-en-17-one (DHA). Endocrinolgy **97**, 418 (1975).
TAKEDA, M., YAMAMURA, H., OHGA, Y.: Phosphoprotein kinases associated with rat liver chromatin. Biochem. biophys. Res. Commun. **42**, 103 (1971).
TATA, J. R.: Membrane phospholipid synthesis and the action of hormones. Nature (Lond.) **213**, 566 (1967).
TATO, F., GANDINI, D. A., TOCCHINI-VALENTINI, S. P.: Major DNA polymerases com-

mon to different Xenopus laevis cell types. Proc. nat. Acad. Sci. (Wash.) 71, 3706 (1974).

TENG, C. S., TENG, C. T., ALLFREY, V. G.: Studies on nuclear acidic proteins. Evidence for their phosphorylation, tissue specificity, selective binding to DNA and stimulatory effects on transcription. J. biol. Chem. 246, 3597 (1971).

TETTENBORN, U., DOFUKU, R., OHNO, S.: Noninducible phenotype exhibited by a proportion of female mice heterozygous for the X-linked testicular feminization mutation. Nature (Lond.) [New Biol.] 234, 37 (1971).

THIEULANT, M. L., MERCIER, L., SAMPAREZ, S., JUOAN, P.: Dihydrotestosterone protein binding in the cytosol of rat center hypophysis *in vitro;* evidence for a specific receptor. J. Steroid Biochem. 6, 1257 (1975).

THIEULANT, M. L., NGUYEN, C. T., SAMPAREZ, S., JOUAN, P.: Increase by castration of testosterone A-ring reductase in the anterior hypophysis of male rats. Biochimie 55, 991 (1973).

THOMAS, P. Z.: Metabolism of andros-4-ene-3,17-dione-^{14}C by rabbit skeletal muscle *in vitro.* J. biol. Chem. 243, 6110 (1968).

TINDALL, D. J., FRENCH, F. S., NAYFEH, S. N.: Androgen uptake and binding in rat epididymal nuclei, *in vivo.* Biochem. biophys. Res. Commun. 49, 1391 (1972).

TINDALL, D. J., HANSSON, V., MCLEAN, W. S., RITZÉN, E. M., NAYFEH, S. N., FRENCH, F. S.: Androgen-binding protein in rat epididymis: properties of a cytoplasmic receptor for androgens similar to the receptor in rat prostate and different from androgen-binding protein (ABP). Molec. cell. Endocr. 3, 83 (1975).

TOFT, D.: The interactions of uterine estrogen receptors with DNA, J. Steroid Biochem. 3, 515 (1972).

TOFT, D., GORSKI, J.: A receptor molecule for estrogens: isolation from the rat uterus and preliminary characterization. Proc. nat. Acad. Sci. (Wash.) 55, 1574 (1966).

TOMINO, S., PAIGEN, K.: Egasyn, a protein complexed with microsomal β-glucuronidase. J. biol. Chem. 250, 1146 (1975).

TÓTH, M., ZAKÁR, T.: Extraction of a secretory protein from the tissues of the seminal vesicle of the rat. Acta biochim. biophys. Acad. Sci. hung. 6, 231 (1971).

TRAPP, G. A., SEAL, U. S., DOE, R. P.: A ligand column for the purification of steroid binding proteins. Steroids 18, 421 (1971).

TREMBLAY, R. R., BEITINS, I. Z., KOWARSKI, A., MIGEON, C. J.: Measurement of plasma DHT by competitive protein-binding analysis. Steroids 16, 29 (1970).

TREMBLAY, R. R., FOREST, M. G., SHALF, J., MARTEL, J. G., KOWARSKI, A., MIGEON, C. J.: Studies on the dynamics of plasma androgens and on the origin of dihydrotesterosterone in dogs. Endocrinology 91, 556 (1972).

TSCHOPP, E.: Untersuchungen über die Wirkung der männlichen Sexualhormone und iher Derivate. Arch. int. Pharmacodyn. Ther. 52, 381 (1936).

TUOHIMAA, P., NIEMI, M.: The effect of testosterone on cell renewal and mitotic cycles in sex accessory glands of castrated mice. Acta endocr. (Kbh.) 58, 696 (1968).

TUOHIMAA, P., NIEMI, M.: Uptake of sex steroids by the hypothalamus and anterior pituitary of pre- and neo-natal rats. Acta endocr. (Kbh.) 71, 37 (1972).

TUOHIMAA, P., NIEMI, M.: Cell renewal and mitogenic activity of testosterone in male sex accessory glands. Male accessory sex organs, p. 327, D. Brandes (ed.). New York: Academic Press 1974.

TVETER, K. J.: Some aspects of the pathogenesis of prostatic hyperplasia. Acta path. microbiol. scand. Sect. A. Suppl. 248, 167 (1974).

TVETER, K. J., ATTRAMADAL, A.: Selective uptake of radioactivity in rat ventral prostate following administration of testosterone-1,2-^{3}H. Acta endocr. (Kbh.) 59, 218 (1968).

TVETER, K. J., ATTRAMADAL, A.: Autoradiographic localisation of androgen in the rat ventral prostate. Endocrinology 85, 350 (1969).

TYMOCZKO, J. L., LIAO, S.: Retention of an androgen-protein complex by nuclear chromatin aggregates: heat labile factors. Biochim. biophys. Acta 252, 607 (1971).

UNHJEM, O.: Partial separation of a 3α-ketosteroid oxidoreductase and an androgen binding substance present in rat ventral prostate cytoplasm. Acta endocr. (Kbh.) **65**, 525 (1970a).

UNHJEM, O.: Metabolism and binding of oestradiol-17β by rat ventral prostate *in vitro*. Research on steroids, vol. IV, p. 139, M. Finkelstein, A. Klopper, C. Conti, C. Cassano (eds.). Oxford: Pergamon Press 1970 b.

UNHJEM, O., TVETER, K. J.: Localization of an androgen binding substance from the rat ventral prostate. Acta endocr. (Kbh.) **60**, 571 (1969).

UNHJEM, O., TVETER, K. J., AAKVAAG, A.: Preliminary characterization of an androgen-macromolecular complex from the rat ventral prostate. Acta endocr. (Kbh.) **62**, 153 (1969).

VALLADARES, L., MINGUELL, J.: Characterization of a nuclear receptor for testosterone in rat bone marrow. Steroids **25**, 13 (1975).

VANNIER, B., RAYNAUD, J. P., Effect of estrogen plasma binding on sexual differentiation of the rat foetus. Molec. Cell Endocr. **3**, 323 (1975).

VERHOEVEN, G.: Androgen binding in prostate and kidney as studied by ammonium sulphate precipitation. J. Steroid Biochem. **5**, 346 (1975).

VERMEULEN, A., STOÏCA, T., VERDONCK, L.: The apparent free testosterone concentration, an index of androgenicity. J. clin. Endocr. **33**, 759 (1971).

VERMEULEN, A., VERDONCK, L.: Studies on the binding of testosterone to human plasma. Steroids **11**, 609 (1968).

VILLEE, C. A., GRIGORESCU, A., REDDY, P. R. K.: Androgen regulation of RNA synthesis in target tissues. J. Steroid Biochem. **6**, 561 (1975).

VITTEK, J., ALTMAN, K., GORDON, G. G., SOUTHREN, L. A.: The metabolism of 7α-^{3}H-testosterone by rat mandibular bone. Endocrinology **94**, 325 (1974).

VLIET, P. C. VAN DER, LEVINE, A. J.: DNA-binding proteins specific for cells infected by adenovirus. Nature (Lond.) [New Biol.] **246**, 170 (1973).

VOGEL, T., SINGER, M. F.: Interaction of f_1 histone with superhelical DNA. Proc. nat. Acad. Sci. (Wash.) 72, 2597 (1975).

VOGT, V. M.: Purification and further properties of single-strand specific nuclease from *Aspergillus oryzae*. Europ. J. Biochem. **33**, 192 (1973).

VOIGT, K., HSIA, S. L.: The antiandrogenic action of 4-androsten-3-one-17β-carboxylic acid and its methyl ester on hamster flank organ. Endocrinology **92**, 1216 (1973).

VREEBURG, J. T. M.: Distribution of testosterone and 5α-dihydrotestosterone in rat epididymis and their concentration in efferent duct fluid. J. Endocr. **67**, 203 (1975).

WAGNER, R. K., JUNGBLUT, P. W.: Differentiation between steroid hormone receptors, CBG and SBG in human target organ extracts by a single-step assay. Molec. cell. Endocr. **4**, 13 (1976).

WALSH, P. C., MADDEN, J. D., HARROD, M. J., GOLDSTEIN, J. L., MCDONALD, P. C., WILSON, J. D.: Familial incomplete male pseudohermaphroditism, type 2. Decreased dihydrotestosterone formation in pseudovaginal perineoscrotal hypospadias. New Engl. J. Med. **291**, 944 (1974).

WANG, J. C.: Interactions between DNA and an *Escherichia coli* protein ω. J. molec. Biol. **55**, 523 (1971).

WANKA, F., SCHRAUWEN, P. J. A.: Selective inhibition by cycloheximide of ribosomal RNA synthesis in chlorella. Biochim. biophys. Acta **254**, 237 (1971).

WEINER, A. L., OFNER, P., SWEENEY, E. A.: Metabolism of testosterone-4-^{14}C by the canine submaxillary gland *in vivo*. Endocrinology **87**, 406 (1970).

WEINTRAUB, H.: Release of discrete subunits after nuclease and trypsin digestion of chromatin. Proc. nat. Acad. Sci. (Wash.) 72, 7212 (1975).

WEISZ, J., GIBBS, C.: Metabolites of testosterone in the brain of the newborn female rat after the injection of tritiated testosterone. Neuroendocrinology **14**, 72 (1974).

WELLS, L. J.: Hormones and sexual differeniation in placental mammals. Arch. Anat. micr. Morph. exp. **39**, 499 (1950).

WESSELLS, N. K., SPOONER, B. S., ASH, J. F., BRADLEY, M. O., LUDUENA, M. A., TAYLOR, E. L., WRENN J. T., YAMANADA, K. M.: Microfilaments in cellular and developmental processes. Science **171**, 135 (1971).

WHALEN, R. E., LUTTGE, W. G.: Testosterone, androstenedione and dihydrotestosterone – effects on mating behaviour of male rats. Horm. Behav. 2, 117 (1971a).

WHALEN, R. E., LUTTGE, W. G.: Perinatal administration of dihydrotestosterone to female rats and the development of reproductive function. Endocrinology 89, 1320 (1971b).

WHALEN, R. E., RESEK, D. L.: Localization of androgenic metabolites in the brain of rats administered testosterone or dihydrotestosterone. Steroids 20, 717 (1972).

WHALEY, W. G., DANWALDER, M., KEPHART, J. E.: Golgi apparatus. Influence on cell surfaces. Science 175, 596 (1972).

WICKS, W. D., KENNEY, F. T.: RNA synthesis in rat seminal vesicle; stimulation by testosterone. Science 144, 1346 (1964).

WIEST, W. G.: Progesterone and 20α-hydroxypregn-4-en-3-one in plasma, ovaries and uteri during pregnancy in the rat. Endocrinology 87, 43 (1970).

WILKINS, M. H. T.: Physical studies of the molecular structure of deoxyribosenucleic acid and nucleoprotein. Cold Spr. Harb. Symp. quant. Biol. 21, 75 (1956).

WILKINSON, D. S., CIHAK, S., PITOT, H. C.: Inhibition of ribosomal ribonucleic acid maturation in rat liver by 5-fluoro-orotic acid resulting in selective labelling of cytoplasmic messenger ribonucleic acid. J. biol. Chem. 246, 6418 (1971).

WILLEMS, M., PENMAN, M., PENMAN, S.: The regulation of RNA synthesis and processing in the nucleolus during inhibition of protein synthesis. J. Cell Biol. 41, 177 (1969).

WILLIAMS-ASHMAN, H. G.: Changes in the enzymatic constitution of the ventral prostate gland induced by androgenic hormones. Endocrinology 54, 121 (1954).

WILLIAMS-ASHMAN, H. G., BANKS, J.: The synthesis and degradation of citric acid by ventral prostate tissue. J. biol. Chem. 208, 337 (1954).

WILLIAMS-ASHMAN, H. G., REDDI, A. H.: Actions of vertebrate sex hormones. Ann. Rev. Physiol. 33, 31 (1971).

WILLIAMS-ASHMAN, H. G., SHIMAZAKI, J.: Some metabolic and morphogenetic effects of androgens on normal and neoplastic prostate. Endogenous factors influencing host-tumour balance, part III, p. 31, R. W. Wissler, T. L. Lao and S. Woods (eds.). Chicago: University of Chicago Press 1967.

WILLIAMSON, R., CROSSLEY, J., HUMPHRIES, S.: Translation of mouse globin messenger RNA from which the poly(adenylic acid) sequence has been removed. Biochemistry 13, 703 (1974).

WILLIER, B. H.: The embryonic development of sex. Sex and internal secretions, 2nd edition, p. 281, E. Allen, C. H. Danforth, E. A. Doisy (eds.). Baltimore: Williams and Wilkins Co., 1939.

WILLMER, E. N.: Steroids and cell surfaces. Biol. Rev. 134, 368 (1961).

WILSON, J. D.: Testosterone uptake by the urogenital tract of the rabbit embryo. Endocrinology 92, 1192 (1973).

WILSON, J. D.: Dihydrotestosterone formation in cultured human fibroblasts. J. biol. Chem. 250, 3498 (1975).

WILSON, J. D., GLOYNA, R. E.: The intranuclear metabolism of testosterone in the accessory organs of reproduction. Recent Progr. Horm. Res. 26, 309 (1970).

WILSON, J. D., GOLDSTEIN, J. L.: Evidence for increased cytoplasmic androgen binding in the submandibular gland of the mouse with testicular feminization. J. biol. Chem. 247, 7342 (1972).

WILSON, J. D., LASNITZKI, I.: Dihydrotestosterone formation in the fetal tissue of the rabbit and rat. Endocrinology 89, 659 (1971).

WILSON, J. D., LOEB, P. M.: Intranuclear localization of testosterone-1,2-H^3 in the preen gland of the duck. J. clin. Invest. 44, 1111 (1965).

WILSON, J. D., SIITERI, P. K.: The role of steroid hormones in sexual differentiation. Excerpta Medica Int. Congr. Ser. 273, 1051 (1973).

WILSON, J. D., WALKER, J. D.: The conversion of testosterone to 5α-androstan-17β-ol-3-one (DHT) by skin slices of man. J. clin. Invest. 48, 371 (1969).

WILSON, M. J., AHMED, K.: Localization of protein phosphokinase activities in the

nucleolus distinct from extra-nucleolar regions in rat ventral prostate nuclei. Exp. Cell Res. **93**, 261 (1975).
WOLFF, M. E., FELDMAN, D., CATSOULACOS, P., FUNDER, J. W., HANCOCK, C., AMANO, Y., EDELMAN I. S.: Steroidal 21-diazoketones: photogenerated corticosteroid receptor labels. Biochemistry **14**, 1750 (1975).
WOLFF, M. E., KASUYA, Y.: C-3-oxygenation of 17-methyl-5-androstan-17-ol by rabbit liver homogenate. J. med. Chem. **15**, 87 (1972).
WOODCOCK, C. L. F.: Ultrastructure of inactive chromatin. J. Cell Biol. **59**, 368 (1973).
WOTIZ, H. H., DAVIS, J. W., LEMON, H. M.: Steroid biosynthesis by surviving testicular tumour tissue. J. biol. Chem. **216**, 677 (1955).
YAMAGUCHI, K., MINESITA, T., KASAI, H., KURACHI, K., MATSUMOTO, K.: Detection of 17β-hydroxy-5α-androstan-3-one-^{3}H following administration of testosterone-^{3}H in the androgen dependent mouse tumour, Shionogi carcinoma 115. Steroids **17**, 345 (1971).
YAMAMOTO, K. R.: To be published; personal communication to the author (1976).
YAMAMOTO, K. R., ALBERTS, B. M.: On the specificity of the binding of the estradiol receptor to DNA. J. biol. Chem. **249**, 7076 (1974).
YAMAMOTO, K. R., ALBERTS, B. M.: The interaction of the estradiol-receptor protein with the genome: an argument for the existence of undetected specific sites. Cell **4**, 301 (1975).
YAMAMOTO, K. R., STAMPFER, M. R., TOMKINS, G. M.: Receptors from glucocorticoid-sensitive lymphoma cells and two classes of insensitive clones; physical and DNA-binding properties. Proc. nat. Acad. Sci. (Wash.) 71, 3901 (1974).
YEH, W-S., MCGUIRE, M., CENTER, M. S., CONSIGLI, R. A.: Partial purification and properties of a DNA-binding protein from nuclei of cells infected with polyoma virus. Biochim. biophys. Acta **418**, 266 (1976).
YU, J. Y-L., MARQUARDT, R. R.: Synergism of testosterone and estradiol in the development and function of the magnum from the immature chicken (*Gallus domesticus*) oviduct. Endocrinology **92**, 563 (1973).
ZANATI, G., WOLFF, M. E.: Synthesis of androgenic-anabolic nonsteroid. J. med. Chem. **16**, 90 (1973).
ZÖLLNER, E. J., HEICKE, B., ZAHN, R. K.: Human serum deoxyribonuclease assays in [^{3}H] DNA-polyacrylamide gels without staining artifacts. Analyt. Biochem. 58, 71 (1974).

Subject Index

Page numbers in **bold face** refer to principle references.

Monographs on Endocrinology

Editors:
F. Gross, M. M. Grumbach, A. Labhart, M. B. Lipsett, T. Mann, L. T. Samuels, J. Zander

Vol. 1
OHNO, S.: **Sex Chromosomes and Sex-linked Genes**
With 33 figures. X, 192 pages. 1967.

Vol. 2
EIK-NES, K. B., and E. C. HORNING: **Gas Phase Chromatography of Steroids**
With 85 figures. XV, 382 pages. 1968.

Vol. 3
SULMAN, F. G.: **Hypothalamic Control of Lactation**
With 58 figures. XII, 235 pages. 1970.

Vol. 4
WESTPHAL, U.: **Steroid-Protein Interactions**
With 144 figures. XIII, 567 pages. 1971.

Vol. 5
MÜLLER, J.: **Regulation of Aldosterone Biosynthesis**
With 19 figures. VII, 139 pages. 1971.

Vol. 6
FEDERLIN, K.: **Immunopathology of Insulin. Clinical and Experimental Studies**
With 53 figures. XIV, 185 pages. 1971.

Vol. 7
HORTON, E. W.: **Prostaglandins**
With 97 figures. X, 197 pages. 1972.

Vol. 8
GURPIDE, E.: **Tracer Methods in Hormone Research**
With 35 figures. XI, 188 pages. 1975.

Vol. 9
MANCINI, R. E.: **Immunologic Aspects of Testicular Function**
With 36 figures. IX, 114 pages. 1976.

Springer-Verlag New York Heidelberg Berlin

MIX
Papier aus verantwortungsvollen Quellen
Paper from responsible sources
FSC® C105338

If you have any concerns about our products,
you can contact us on
ProductSafety@springernature.com

In case Publisher is established outside the EU,
the EU authorized representative is:
Springer Nature Customer Service Center GmbH
Europaplatz 3, 69115 Heidelberg, Germany

Printed by Libri Plureos GmbH
in Hamburg, Germany